PROTOPLASMATOLOGIA
HANDBUCH
DER PROTOPLASMAFORSCHUNG

HERAUSGEGEBEN VON

L. V. HEILBRUNN UND F. WEBER
PHILADELPHIA   GRAZ

MITHERAUSGEBER

W. H. ARISZ-GRONINGEN · H. BAUER-WILHELMSHAVEN · J. BRACHET-BRUXELLES · H. G. CALLAN-ST. ANDREWS · R. COLLANDER-HELSINKI · K. DAN-TOKYO · E. FAURÉ-FREMIET-PARIS · A. FREY-WYSSLING-ZÜRICH · L. GEITLER-WIEN · K. HÖFLER-WIEN · M. H. JACOBS-PHILADELPHIA · D. MAZIA-BERKELEY · A. MONROY-PALERMO · J. RUNNSTRÖM-STOCKHOLM · W. J. SCHMIDT-GIESSEN · S. STRUGGER-MÜNSTER

BAND III

CYTOPLASMA — ORGANELLEN

A

CHONDRIOSOMEN, MIKROSOMEN, SPHAEROSOMEN

4

CHEMISTRY AND PHYSIOLOGY OF MITOCHONDRIA AND MICROSOMES

SPRINGER-VERLAG WIEN GMBH 1954

CHEMISTRY AND PHYSIOLOGY OF MITOCHONDRIA AND MICROSOMES

BY

OLOV LINDBERG AND LARS ERNSTER

WENNER-GREN'S INSTITUTE
STOCKHOLM

WENNER-GREN'S INSTITUTE
STOCKHOLM

WITH 32 FIGURES

SPRINGER-VERLAG WIEN GMBH 1954

ISBN 978-3-662-24409-8   ISBN 978-3-662-26542-0 (eBook)
DOI 10.1007/978-3-662-26542-0

PROTOPLASMATOLOGIA

HANDBUCH
DER PROTOPLASMAFORSCHUNG

IN 14 BÄNDEN

HERAUSGEGEBEN VON

L. V. HEILBRUNN UND **F. WEBER**
PHILADELPHIA   GRAZ

MITHERAUSGEBER

W. H. ARISZ-GRONINGEN · H. BAUER-WILHELMSHAVEN · J. BRACHET-BRUXELLES · H. G. CALLAN-ST. ANDREWS · R. COLLANDER-HELSINKI · K. DAN-TOKYO · E. FAURÉ-FREMIET-PARIS · A. FREY-WYSSLING-ZÜRICH · L. GEITLER-WIEN · K. HÖFLER-WIEN · M. H. JACOBS-PHILADELPHIA · D. MAZIA-BERKELEY · A. MONROY-PALERMO · J. RUNNSTRÖM-STOCKHOLM · W. J. SCHMIDT-GIESSEN · S. STRUGGER-MÜNSTER

Mitarbeiter

TH. F. ANDERSON, Philadelphia: W. H. ARISZ, Groningen: R. BIEBL, Wien: G. BLUM, Freiburg, Schweiz: H. L. BOOIJ, Leiden: J. BRACHET, Bruxelles: H. G. BUNGENBERG DE JONG, Leiden: H. G. CALLAN, St. Andrews: K. DAN, Tokyo: P. DANGEARD, Bordeaux: H. DRAWERT, Berlin-Dahlem: R. DURYEE, Washington: L. ERNSTER, Stockholm: J. EYMÉ, Bordeaux: E. FAURÉ-FREMIET, Paris: A. FREY-WYSSLING, Zürich: L. GEITLER, Wien; H. GERM, Wien: E. N. HARVEY, Princeton: L. V. HEILBRUNN, Philadelphia: K. HÖFLER, Wien; L. HOFMEISTER, Wien: P. HUBER, Zürich: A. HUGHES, Cambridge, England: M. H. JACOBS, Philadelphia: J. D. JUDAH, London: N. KAMIYA, Osaka: J. W. KELLY, Richmond: J. A. KITCHING, Bristol, England: C. A. KNIGHT, Berkeley: E. KUPKA, Graz; E. KÜSTER †, Gießen: A. I. LANSING, St. Louis; P. G. LE FEVRE, Washington: O. LINDBERG, Stockholm: S. E. LURIA, Urbana: MacCARDLE, Bethesda: H. MARQUARDT, Freiburg i. Br.: D. MAZIA, Berkeley: A. D. J. MEEUSE, Pretoria: G. PALADE, New York: A. K. PARPART, Princeton: E. S. PERNER, Münster: H. H. PFEIFFER, Bremen: G. PIEKARSKI, Bonn-Venusberg: A. POLICARD, Paris: E. PONDER, Mineola: LOTTE REUTER, Wien: P. RIESER, Philadelphia: A. ROTHSTEIN, Rochester: H. SCHINDLER, Wien: W. J. SCHMIDT, Gießen: J. SMALL, Belfast: K. M. SMITH, Cambridge, England: T. M. SONNEBORN, Bloomington: W. C. SPECTOR, London: W. M. STANLEY, Berkeley: H. STAUDINGER und MAGDA STAUDINGER, Freiburg i. Br.: S. STRUGGER, Münster: A. TYLER, Cambridge, England: H. ULLRICH, Stuttgart-Berg: K. UMRATH, Graz: B. WADA, Tokyo: V. WARTIOVAARA, Helsinki: F. WEBER, Graz: F. WIERCINSKI, Philadelphia: W. L. WILSON, Burlington: K. ZEIGER, Hamburg: und andere

130. 3. 54. 20. B.

Disposition

Chemistry and Physiology of Mitochondria and Microsomes

By

OLOV LINDBERG and LARS ERNSTER

Wenner-Gren's Institute, Stockholm, Sweden

With 32 Figures

Contents

Abbreviations

AMP = adenylic acid (adenosine-5-phosphoric acid)
ADP = adenosinediphosphoric acid
ATP = adenosinetriphosphoric acid
DPN = diphosphopyridine nucleotide
TPN = triphosphopyridine nucleotide
FAD = flavine adenine dinucleotide

CoA = coenzyme   A*   (adenosine-[3-phosphoryl]-5-pyrophosphoryl-pantothenyl-
   β-alanylaminoethylmercaptide)
RNA = ribonucleic acid
DNA = desoxyribonucleic acid

Introduction

The study of cytoplasmic particles is today one of the most active fields of cytological research. We have attempted to describe the main features of the historical development in this field in a brief survey (Fig. 1), rather than to present exhaustive lists of data. Further historical aspects are, however, touched upon where practicable in the individual sections.

We wish especially to direct the attention of the reader to the convergence of cytological and enzymological lines of investigation which has become increasingly evident during the past decade in the study of the cytoplasmic particles. It has been possible to identify the mitochondria as the centres of cell respiration and of the distribution of metabolic energy within the cell. The microsomes, on the other hand, have been found to be the bearers of the cytoplasmic nucleoproteins and hence participate in protein synthesis and in the maintenance of the genetic continuity of the cytoplasm. Thus the cytologist has acquired for the first time an enzymological basis for the study of the structural elements of the cell, while the enzymologist has gained access to morphological data concerning his enzyme systems. In this lies the fundamental interest of our field of research.

The present monograph is designed to provide a survey of the chemistry and the physiology of the cytoplasmic particles. It will unfortunately be impossible to cover, even in a first approximation, the literature of our subject, which is enormous in quantity and highly varied in mode of approach. On the other hand, there has not yet been published, to our knowledge, any review in which it has been attempted to cover both the chemical and the physiological aspects of the cytoplasmic particles. We have therefore considered it our main task to collect together results from different lines of investigation, rather than to enter upon a detailed treatment of each line.

Our work has been greatly facilitated by the fact that there are already a considerable number of reviews which cover the various specialized branches. While we have had much recourse to these, we have nevertheless found it desirable, in order to fulfill our above-mentioned purpose, to write each section of the monograph with especial thought to the reader whose speciality is discussed in *other* sections.

In this endeavour we have often been confronted by the problem of balancing the content of a chapter in such a manner that it will be comprehensible to the non-specialist without appearing uninteresting and

---

* In formulae the symbol R_{CoA} SH will be used. Analogously, lipothiamide-pyrophosphate and glutathione will be abbreviated as R_{LTPP}-SH and R_{glut}-SH, respectively.

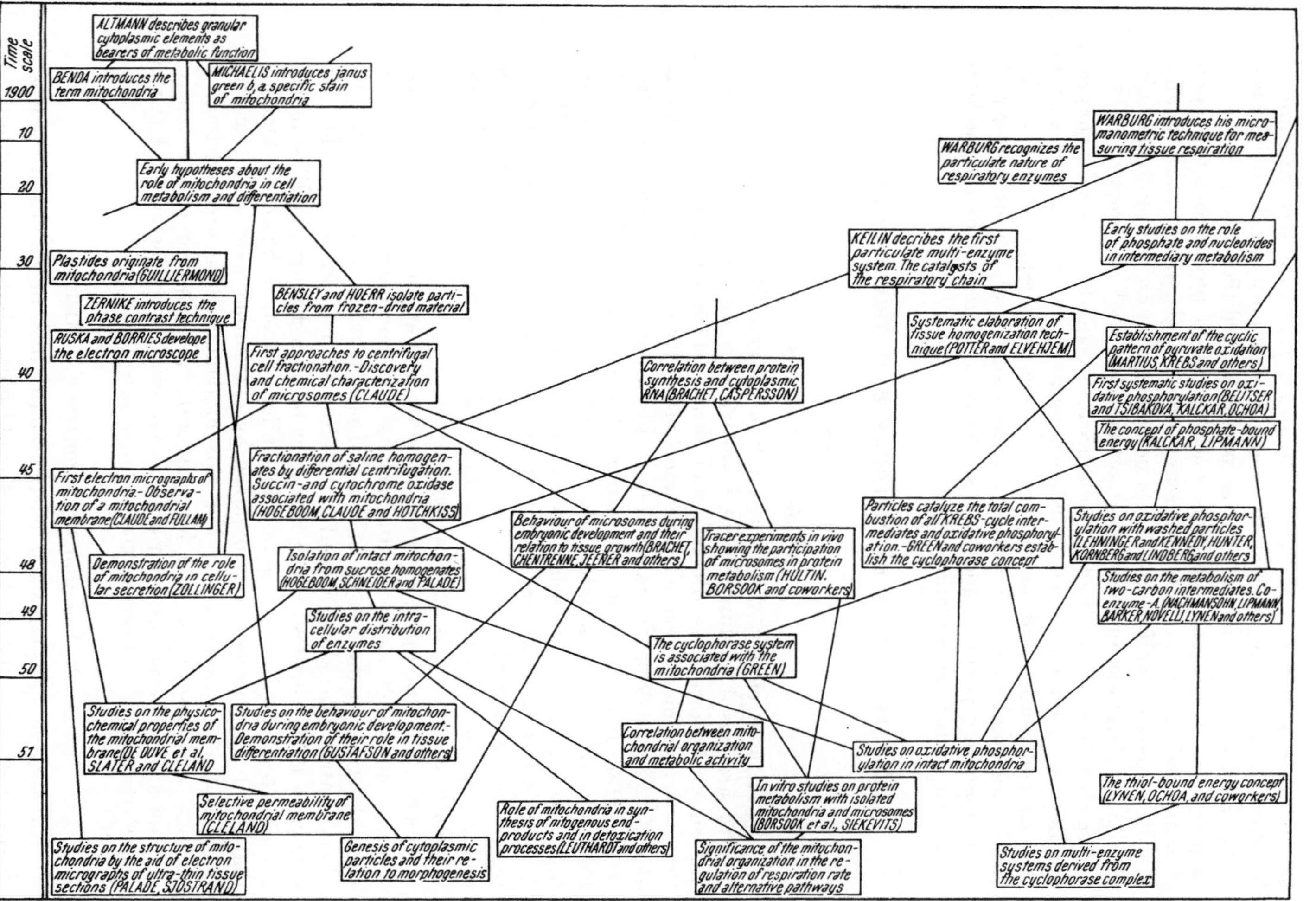

Fig. 1. Historical survey of research trends in the study of cytoplasmic particles

naïve to the specialist. We are well aware that any such attempt is foredoomed to some degree of failure, and we can only beg the indulgence of the reader in this matter. It was particularly difficult to prevent the chapter entitled "Enzymic processes concerned in the physiological function of mitochondria" from exceeding reasonable bounds, since many details, of little apparent significance, had to be included in order to provide links with certain sections in the chapter on mitochondrial physiology.

We therefore thought it advisable to treat the said chapter on the enzyme complement of the mitochondria independently of the two chapters on the chemical constitution and the physiology of the cytoplasmic particles. The three main chapters are preceded by a short survey of the morphology of the cytoplasmic particles and by a similarly brief review of the methods employed in the investigations.

Short Survey of Methods and Morphology

A. Methods

1. Histological methods

The methods employed for the study of mitochondria and microsomes *in situ* are based exclusively on histological techniques, such as fixing and staining procedures (Bensley and Bensley 1938, Cowdry 1943, Lillie 1948), electron microscopy (Claude and Fullam 1946, Mühlethaler, Müller and Zollinger 1950, Sjöstrand 1951), phase-contrast microscopy (Zollinger 1950) and ultraviolet microscopy (Caspersson 1950, Opie and Lavin 1946). A technique especially valuable in the study of unicellular organisms and larvae, has proved to be centrifugal microscopy (Harvey 1937). An account of these methods would be more in place in a morphological review and is beyond the scope of this book. For further information on the application of these methods to the study of cytoplasmic particles the references above should be consulted.

Some of the techniques, however, are also used and are in fact indispensible in studies of isolated mitochondria and microsomes. Perhaps the most important of these is the staining of mitochondria with Janus green B (diethylsafranineazodimethylaniline hydrochloride), introduced by Michaelis in 1900. This test, which is specific for mitochondria, and has indeed actually been used as a criterion for mitochondria, is based on the fact that the dye is specifically re-oxidized by cytochrome oxidase, an enzyme found only in mitochondria. For this explanation we are indebted to Lazarow and associates, who have recently published a comprehensive account of their work (Lazarow and Cooperstein 1953, Cooperstein, Lazarow and Patterson 1953, Cooperstein and Lazarow 1953).

2. Isolation of cytoplasmic particles

The isolation of cytoplasmic particles dates back to the year 1934, when Bensley and Hoerr reported the isolation of mitochondria from frozen-dried material. This initial success was followed by a development of the differential centrifugation technique, in which Lazarow (1943) and Claude (1943a) may be named as the pioneers.

These first experiments in cell fractionation were made with tissues homogenized in isotonic salt solutions and often gave non-uniform, contaminated and damaged particles. In 1948, Hogeboom, Schneider and Palade found that a preparation of mitochondria, uniform and fairly unaffected in the morphological and cytological sense, could be obtained if the tissue was homogenized in 0.88 M sucrose solution. This is the main procedure employed today in mitochondrial preparations.

A number of more or less extensive modifications have subsequently been made in the basic method. These have recently been critically examined by Schneider and Hogeboom (1951). An important variant of the sucrose method, now almost generally accepted, consists in an alteration of the sucrose concentration from 0.88 to 0.25 M (isotonic). This renders the preparation easier, in that smaller centrifugal fields are required for spinning down the particles. Furthermore it eliminates the inhibitory effect of high sucrose concentrations on certain enzyme systems and thus facilitates biochemical studies (Schneider and Hogeboom 1950). An important precaution is to provide the sucrose solution with such a buffering capacity in the pH range 7.5–8.0, as will compensate for the fall in pH which usually occurs in the homogenization of tissues. This may be achieved by addition of a small amount of alkali, or better, of β-glycerophosphate, which has a relatively good buffering power in the required pH range and has the same protective action on the mitochondria as sucrose (Berthet and de Duve 1952).

The isolation procedure in the form now most frequently employed is as follows: The tissue is removed from the animal as quickly as possible, cooled in about 9 volumes of ice-cold sucrose solution and cut into small pieces. All subsequent operations must be carried out at between 0° and 5° C. The pieces of tissue are stirred gently for three minutes in three new lots of sucrose solution and are then homogenized in a Potter-Elvehjem homogenizer (Potter and Elvehjem 1936). If it is thought desirable, the homogenate may be filtered through gauze to remove any coarse particles. Unbroken cells, nuclei and cell débris are removed by low-speed centrifugation (c. 1000 g) for three minutes. The supernatant is then spun at a high rate (c. 14,000 g for isotonic sucrose) for 15 to 20 minutes. The mitochondria are obtained as a pellet and may be washed by careful resuspension in sucrose and high-speed centrifugation. For the isolation of microsome material, the clear or almost clear supernatant is spun at 50,000 to 80,000 g for one hour.

Difficulties may arise in the homogenization of tissues of tough consistency such as heart muscle, or of cell suspensions such as fluid ascites tumours. In such cases the use of quartz sand in the homogenization has proved to be advantageous. The homogenization of tough tissues is carried out in a mortar (Slater 1950 c) and that of fluid tissues by shaking the suspension of cells and sand in a long, thick-walled test tube (Lindberg, Ljunggren, Ernster and Révész 1953). In order to improve the yield of mitochondria, thorough elution of the sand-containing homogenate must be performed. Three to four grindings with fresh quantities of sucrose solution are usually necessary.

It has recently been reported by Slater and Cleland (1952) that if the preparation is carried out in a calcium-free environment, which can be achieved by adding versene to the sucrose solution to a concentration of 0.01 M, the durability of the mitochondria is improved. Mitochondria thus prepared may, for example, be recovered from incubation mixtures and resuspended for further incubations.

3. Quantitative estimation of particles

a) Counting of mitochondria

This procedure is that usually employed in the standardization of isolated mitochondria in cytological investigations, but is equally suitable for biochemical purposes. A method for the counting of isolated mitochondria has recently been systematically developed by Allard, Mathieu, de Lamirande and Cantero (1952). The procedure, which is described in detail by the authors, consists in suspending the mitochondria in 1.8 M sucrose to a concentration of about 0.2%. The number per unit volume is then determined in a Petroff-Hausser bacteria counter.

b) Standardization with reference to protein content

Methods of this type are suitable for biochemical purposes when it is desired, for example, to compare the enzymic activities of different mitochondrial preparations. Harman and Feigelson (1952b) suspended an aliquot of a mitochondrial preparation in strong KOH, boiled for 30 minutes and read the extinction at $280 \, m\mu$. Lindberg, Ljunggren, Ernster and Révész (1953) have adapted the modification by Robinson and Hogden (1940) of the biuret method for the determination of mitochondrial protein. These methods might also be useful for the standardization of microsomes.

4. Biophysical and biochemical techniques

The technique employed in experiments with isolated mitochondria may vary greatly according to the nature of the investigation. It may involve purely physicochemical methods and simple chemical analyses, more or less complicated enzyme incubations or a combination of all of these. As examples of methods that have been used, we may mention the employment of sonic vibrations (Hogeboom and Schneider 1950a) and measurements of light scattering (Cleland 1952) in studies of the osmotic conditions of isolated mitochondria; ultracentrifugation (Hogeboom and Schneider 1951) and electrophoresis (de Lamirande, Allard and Cantero 1953) in studies of the protein components of mitochondria; and the usual manometric technique (cf. Umbreit, Burris and Stauffer (1947) for a general account) with the associated analytical methods in diverse combinations. Use has also been made of paper chromatography in conjunction with the isotope technique, which has been found highly advantageous in the quantitative study of mitochondrial incubations (Lindberg and Ernster 1952a).

B. Morphology

1. Microsomes

Microsomes are strongly basophil, Feulgen-negative, sub-microscopic particles of about 50—150 mμ magnitude (CLAUDE 1940). They are stained by iron haematoxylin (VENDRELY 1950). The basophilic and siderophilic properties of microsomes are explained by their high content of RNA, a matter to which we shall return at a later stage. SCHNEIDER (1947) believes them to be rod-shaped in view of their movement and their birefringence. ZOLLINGER (1950) claims to have observed microsomes under the phase contrast microscope, when the preparation was subjected to prolonged treatment with distilled water or ammonia. According to MONNÉ (1948), who calls them chromidia, they are parts of the fibrillar structure of the ground cytoplasm. In the electron micrographs of CLAUDE and FULLAM (1946), the microsomes appear embedded in a fibrous structure. As the authors state, however, it is impossible to decide whether this is a fixation artefact. According to a report by HUSEBY and BARNUM (1950), electron microscopic observations have failed to demonstrate the existence of a membrane. More recent studies of ZOLLINGER (1951) and of EICHEN-BERGER (1953), which are concerned with the transitional forms between mitochondria and microsomes, confirm this finding. Isolated microsomes have a red-brown colour which, according to BENSLEY (1947), may be attributed to a pigment derived from the oxidation of unsaturated fatty acids.

It has not yet been possible to locate the microsomes within the cell. HERS, BERTHET, BERTHET and DE DUVE (1951) refer, however, to their finding (see below) that the alkaline phosphatase in the kidney is found exclusively in the microsomes, and point out that the Gomori cytochemical method should provide information concerning the intracellular site of these structures. A high concentration of phosphatase in the brush borders of proximal convoluted tubules, found by Gomori's method (cf. MOOG 1946), would therefore suggest that these formations contain aggregations of microsomes.

2. Particles of transitional character

Although the orders of magnitude of mitochondria and microsomes are widely separated, it has long been a matter of interest whether the two groups of cytoplasmic particles are distinct one from the other. This question became very urgent when the technique of differential centrifugation was introduced and when the chemical and enzymic characterization of the particles became an object of intensive study.

Some ground for scepticism was provided by the early studies of CHANTRENNE (1947), who showed that the fractional centrifugation of mouse liver homogenate resulted in not less than five different particulate fractions, distinguishable both chemically and enzymically. This finding led CHANTRENNE to regard as arbitrary the division of cytoplasmic particles into two distinct fractions. This conclusion was challenged by SCHNEIDER and HOGEBOOM (1951) on the ground that the fractionation technique employed by Chantrenne was unsatisfactory. Later, however, the results of Chantrenne were confirmed in some measure by Jeener (1948) who, by fractionation with salt solutions of particulate material isolated in a sucrose medium, obtained sub-fractions differing in size, and in enzymic and chemical constitution. In Jeener's view, we are concerned here with transitional forms between mitochondria and microsomes.

A heterogeneity in the mitochondrial population has recently been

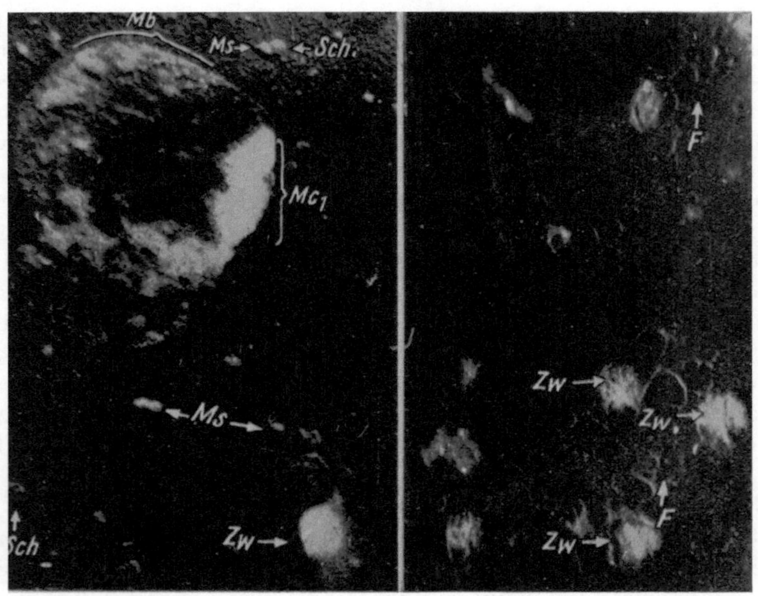

Fig. 2. Formation of mitochondria from microsomes.

Electron micrographs (Cr-shadowed. Magn. 20,000 ✕) on suspension of cytoplasmic particles of mouse kidney. *Left:* Normal. One mitochondrion (*Mc₁*), several microsomes (*Ms*) and one transitional form (*Zw*). *Right:* After destruction of mitochondria by repeated injection of ovalbumin. Mitochondria under regeneration. Several transitional forms, but no mitochondria or microsomes are visible. (*F* = unidentified structures.)
(From Eichenberger 1953.)

postulated by de Duve, Appelmans and Wattiaux (1952), who were able to distinguish between two mitochondrial fractions in respect to their sedimentation rates and enzymic compositions. A similar result has been reported by Laird, Nygaard, Ris and Barton (1953). These last-named authors find that the upper layer ("fluffy layer") in centrifuged mitochondria differs from the close-packed pellet in both RNA content and succinic oxidase activity, which appear to be in inverse ratio in the two fractions. The identity of the fluffy layer with microsomes has been excluded by the admixture of isotopically labelled microsomes. An electron microscopical study of the two mitochondrial fractions showed that they appeared to differ only in size.

The concept of transitional forms between mitochondria and microsomes has in recent years been put on a more satisfactory morphological basis by the Zollinger group. By artificial stimulation of mitochondrial regeneration in kidney tubules, ZOLLINGER (1950) has been able to demonstrate a continuous series of transitional forms between submicroscopic particles and adult mitochondria. These intermediate forms have been followed by EICHENBERGER (1953) under the electron microscope and it has been possible in this manner to obtain valuable information concerning the development of the mitochondrial membrane (Fig. 2).

3. Mitochondria

Mitochondria, in the narrower meaning of the term, are cytoplasmic particles of a magnitude between the limit of resolution of the light microscope and $7\,\mu$ in their largest dimension (COWDRY 1928), and stainable with Janus green B. In the wider sense, the mitochondria are taken to include the transitional forms discussed above, together with particles specialized for secretion, storage, etc. (ZOLLINGER 1950), which are considered to be modified mitochondria. These types are not stained by Janus green.

Recently a succinic oxidase activity has been taken as a criterion of mitochondria in the narrower meaning. When tests for such activity are employed, however, it should be borne in mind that mitochondrial fragments will give positive results. A preferable enzymic criterion is then the capacity for oxidative phosphorylation, which is a more certain indication of intact mitochondria.

a) Form and size

Mitochondria in different forms (and often under different names) occur in every type of living cell so far investigated, both in plants and animals, uni- and multicellular organisms. Their form may vary from one type of cell to another and under different conditions. For a comprehensive review of this subject see ZOLLINGER (1950). In the uninfluenced ("resting") cell, the mitochondria are rod-shaped or granular with a diameter between 0.5 and $1\,\mu$, with 0.2 and $2\,\mu$ as extreme limits, and a maximum length of $7\,\mu$. The diameter within a given cell is constant (BOURNE 1942).

The form of the mitochondria undergoes profound alteration when the cell is subjected to external or internal stimuli. When the cell is absorbing or secreting, or is subjected to hypotonic conditions, the mitochondria swell, become rounded and may assume a volume many times the resting value. In a hypertonic medium, on the other hand, they shrink and the rods are readily converted to angular formations.

A direct relationship between physical state, oxidative activity and form in isolated mitochondria of kidney, liver and muscle is found in the studies of HARMAN (1950a, b) and HARMAN and FEIGELSON (1952a, b, c) with the phase-contrast microscope. These authors were able to distinguish between two main forms: "target" (or if the preparation was made in hypertonic sucrose, "rod") on the one hand and "limbus" and "crescent" on the other. There was also a transitional form between these, described as "spherical dense". These forms have been correlated with different metabolic states, both *in vitro* and *in vivo*.

Ritchie and Hazeltine (1953) have demonstrated profound changes of form in the mitochondria of the mould *Allomyces,* which is said to be an excellent material for studies on mitochondria. These authors are not convinced of a correlation between the physiological state of the organism and the mitochondrial behaviour.

b) Number

The number of mitochondria varies from tissue to tissue. In a normal rat liver cell the number is reported to be about 2500 (Allard, Mathieu, de Lamirande and Cantero 1952). (See Table 1.) In highly active cells the number is relatively high, such as in the motor ganglion cells of the spinal cord (Hartman 1948). An elegant example of this is given by Paul and Sperling (1952) who have demonstrated a high density of mitochondria in the breast muscle of birds which are good fliers, in contrast to the state of affairs in poor fliers, where the complement of mitochondria in the breast muscle is small.

The number of mitochondria may also vary in the same tissue under different physiological conditions. Thus a diminution takes place after gastrulation (Gustafson and Lenique 1952) and in general during growth unattended by differentiation, as in tissue regeneration and in neoplastic tissue (Allard, de Lamirande and Cantero 1952, 1953). An increase in the mitochondrial population has been observed during the differentiation

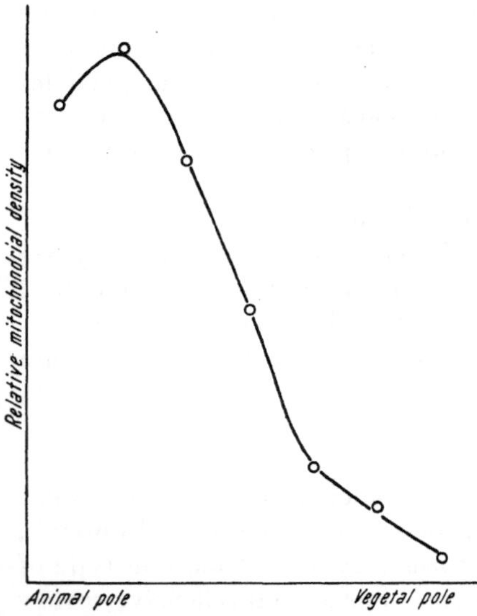

Fig. 3. Gradient of mitochondrial density along the animal-vegetal axis in the larva of the sea urchin, *Psammechinus miliaris,* at the blastula stage. (From Gustafson and Lenicque 1952.)

Table 1. *Quantitative data on mitochondrial density in normal, neoplastic and regenerating mouse liver.*

(After Allard, de Lamirande and Cantero 1953.)

Material	Number of mitochondria per gram wet weight	Number of mitochondria per cell	mg N per mito-chondrion
Normal rat liver	$3.30 \cdot 10^{11}$	2554	$2.23 \cdot 10^{-11}$
Liver tumor	$1.23 \cdot 10^{11}$	1391	$3.08 \cdot 10^{-11}$
Regenerating liver . . { 1 day	$2.42 \cdot 10^{11}$	1800	$2.80 \cdot 10^{-11}$
{ 20 days	$2.58 \cdot 10^{11}$	2150	$2.52 \cdot 10^{-11}$

period in larvae of the sea urchin, where a definite gradient is found along the animal-vegetal axis (GUSTAFSON and LENICQUE 1952) (Fig. 3). An increase in mitochondria has also been observed in vigorously secreting tissues such as secreting tumours (LUDFORD 1948). JUNQUEIRA (1951) has found a direct correlation between the mitochondrial count and the secretory activity in the salivary glands of the mouse.

c) Intracellular distribution

With regard to the intracellular distribution of the mitochondria, it may be stated as a general rule that a polarity in the activity of the cell

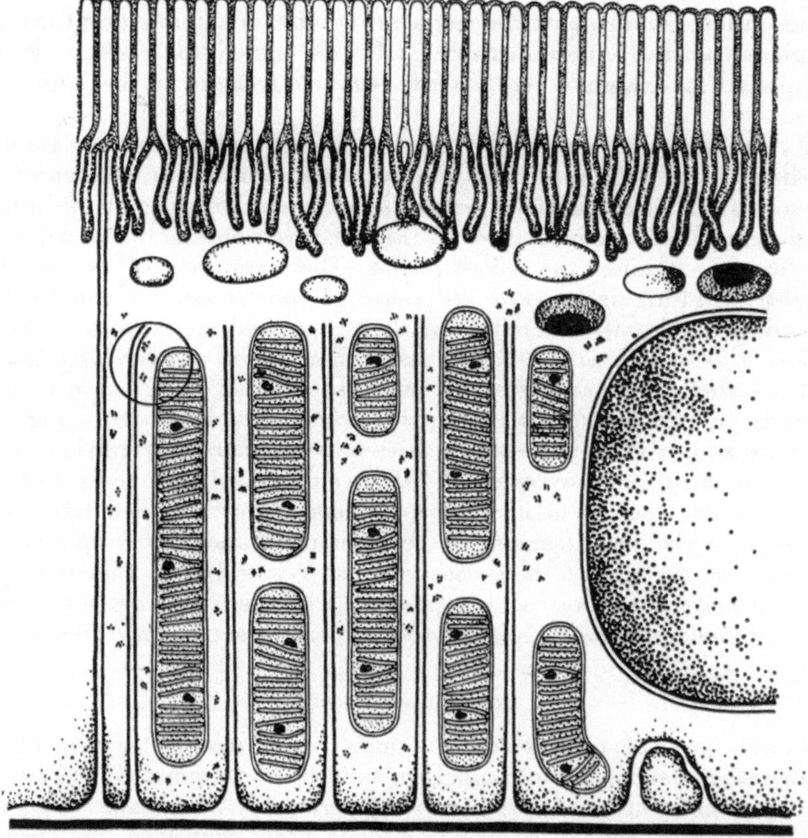

Fig. 4. Schematic presentation of the ultrastructural organization of the proximal tubular cells of the mouse kidney, as reconstructed on the basis of electron micrographs of ultra-thin sections.
In the upper part of the drawing the brush-border cell granules are located and the basal part of the cell is occupied by the mitochondria within intracellular compartments lined by intracellular cytoplasmic membranes. The basal surface of the cell is lined by the basement membrane.
(From SJÖSTRAND and RHODIN 1953.)

is reflected in a concentration of mitochondria. In secretory cells, the mitochondria accumulate basally, in absorbing cells apically (ZOLLINGER 1950). In the proximal convoluted tubules of the kidney, which belong to the latter category, they are said to be fixed in the basal membrane (ZOLLIN-

GER 1950). According to a recent investigation by Sjöstrand and Rhodin (1953), they are inserted one by one or two by two in pockets of the cell membrane, without connection with the basal membrane (Fig. 4). In sperm cells, the well-known accumulation of mitochondria in the neck may be associated with the active rôle of this region in the movement of the cell. The sarcosomes of muscle are intimately bound to the contractile elements (Watanabe and Williams 1951 (Fig. 29, p. 101), Harman and Feigelson 1952 b). In the retinal rods the mitochondria are found proximally at the photo-receptor end and parallel with the long axis (Sjöstrand 1953 b).

An alteration in the polarity of a cell may find expression in the mito-chondrial distribution. Thus extensive redistribution of mitochondria occurs in the cell during mitosis and can be followed in an elegant manner by phase-contrast cinematography (Hughes and Fell 1949). During metaphase for example, the mitochondria congregate in the equatorial plane.

A correlation between active loci in the cell and an accumulation of mitochondria at these loci might provide a basis for an explanation of the movements of mitochondria. Very little is known, however, of the stimuli which release these movements or of the forces responsible in the polarized cell for the fixation of mitochondria. The existence of cytoplasmic structures which might serve as points of anchorage for mitochondria (Heidenhain's "Basalreifen") has not been confirmed in the phase-contrast studies of Zollinger (1950). Nor has a fixation of mitochondria at the basal membrane of the kidney tubule, as observed by Zollinger, been confirmed by the electron micrographs of Sjöstrand (see above). Frédéric and Chèvremont (1952) have used phase-contrast cinematography to follow the development and movements of chick embryo skeletal muscle cultures. They describe, among other things, movements of mitochondria which suggest an exchange of substances between these and the cytoplasm and nucleus. In the light of these observations, it would seem that the mito-chondrial movements are not determined, as suggested by Zollinger, only by cytoplasmic streaming, but may be of a more active character.

d) Structure

On the basis of phase-contrast studies, the internal structure of the mitochondria is supposed to comprise a solid body surrounded by a hydro-philic sol and enclosed in a selectively permeable membrane (Zollinger 1948). The existence of a membrane has been confirmed by electron microscopy (Claude and Fullam 1945, Porter, Claude and Fullam 1945, Claude and Fullam 1946, Dalton, Kahler, Kelly, Lloyd and Striebich 1949, Mühlethaler, Müller and Zollinger 1950, Huseby and Barnum 1950).

Palade (1952) has studied the fine structure of mitochondria in thin ($<$ 1000 Å) sections of a great variety of tissues by the aid of electron microscopy after fixation in osmic acid. All the investigated types of mitochondria have been found to possess a) a limiting membrane, 70–80 Å thick; b) a system of internal ridges (called *cristae mitochondriales*), of

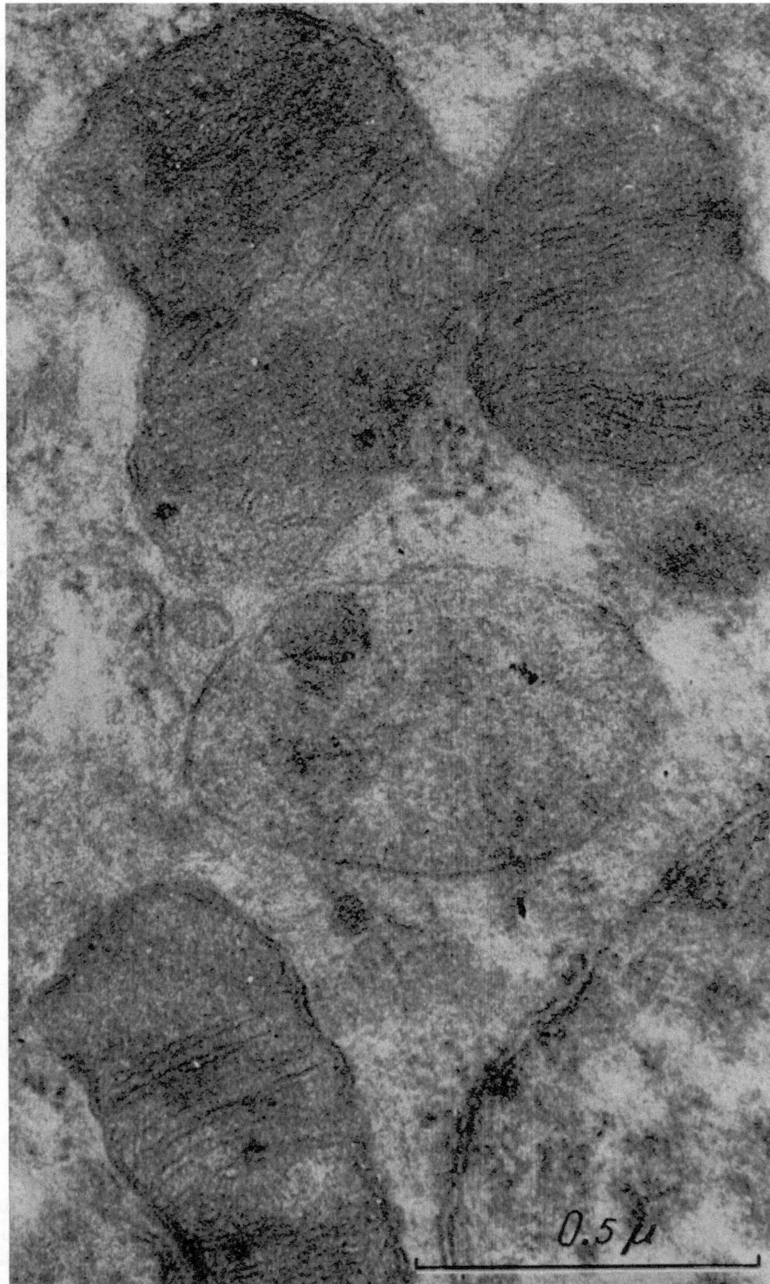

Fig. 5. Mitochondria in a proximal convoluted tubule of the mouse kidney.
Three mitochondria are seen, two in the upper part of the picture, one in the lower left corner. In the centre a cell granule and in the lower right-hand corner a part of a cell nucleus. The mitochondria are surrounded by a 160 Å thick double membrane and in the interior of the mitochondria a system of 160 Å thick double-edged membranes is seen. The thickness of the constituent single membranes is 45 Å and the height of the space in between the two constituent single membranes of the double membranes is 70 Å. — Ultrathin section. Magnification 97,000 ×.
(From SJÖSTRAND and RHODIN 1953.)

a uniform thickness of 180–200 Å, protruding circularly from the inner surface of the membrane, and situated at more or less regular intervals along the long axis of the rod-shaped mitochondrion; and c) a matrix. The finer structure of the membrane could not be visualized at the resolution power obtained. The *cristae,* on the other hand, have been found to have a trilaminar structure with a central layer, 80–100 Å thick, covered on each side by a denser layer of 50–70 Å thickness.

Similar results have recently been reported by Sjöstrand (1953a) and Sjöstrand and Rhodin (1953). These authors succeeded in a considerable elevation of the resolving power by achieving ultra-thin (c. 100 Å) tissue sections. The electron micrographs (Fig. 5) obtained with this technique show a membrane of about 160 Å thickness, consisting of two protein layers, 45 Å each, with an intermediate lipid layer of 70 Å. A large number of transversely oriented interior septa, in structure very similar to the membrane, are probably idential with Palade's *cristae.*

A striated structure of the mitochondrial body has also been described by Glimstedt and Lagerstedt (1953). By electron microscopic investigations of isolated rat liver mitochondria fixed in osmic acid, these authors have

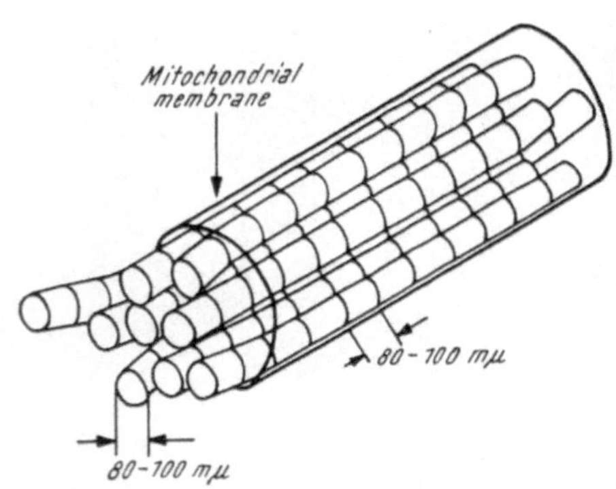

Fig. 6. Schematic drawing showing a combination of observations of isolated mitochondria.
(From Glimstedt and Lagerstedt 1953.)

come to the conclusion that the solid core included in the mitochondrial membrane consists of a cable-like structure of equal-sized granules, 800–1000 Å in diameter. A schematic model of this structure is shown in Fig. 6.

The tendency to recognize differentiated structural elements within the mitochondria may be associated with the fact that a number of enzyme systems are known to retain their particulate character when detached from the mitochondrial entity. Palade points out that the *cristae* are of about the same size as isolated succinicoxidase, and, moreover, that the number of *cristae* per mitochondrion is very high in heart muscle, a tissue especially rich in succinicoxidase.

The phenomenon of "fragmentation" of mitochondria into particulate enzymic components has earlier been emphasized, on a biochemical basis, by Green (1951b), although this author failed to realize the fundamental difference between these fragments and microsomes. Green (1952) also, referring to investigations by Harman (1950), refuses to accept the existence of a mitochondrial membrane as proved. This opinion is, however, hardly maintainable today. A further discussion of this matter follows on p. 26.

Chemical Constitution of Microsomes and Mitochondria

A. Microsomes

A comprehensive review of the chemical constitution of microsomes has been published by SCHNEIDER and HOGEBOOM (1951), on which the following short account is based.

The microsomes constitute a surprisingly large proportion of the cytoplasm, about 20% on the basis of total nitrogen. This figure is somewhat lower for certain neoplasms. The most striking chemical characteristic of the microsomes is their high content of RNA and of lipids. The high proportion of RNA was first observed by CLAUDE (1944, 1946). Liver microsomes contain not less than 50% of the total RNA of the tissue with a concentration, expressed as RNA phosphorus per mg. nitrogen, of 63 μg. as against 27 μg. for the tissue as a whole. The RNA concentration in kidney microsomes is reported to be somewhat lower. Variations in both directions have been observed in different types of neoplastic tissue (PRICE, MILLER, MILLER and WEBER 1949 a, b, 1950). A diminution in the RNA of microsomes during starvation has been reported by VENDRELY (1950).

The total lipid content of the microsomes is about 43% of the dry weight and 21% of the total lipid content of the cytoplasm (CLAUDE 1946, ADA 1949, HUSEBY and BARNUM 1950, LEVIN and CHARGAFF 1952). 30% of the dry weight consists of phosphatides, which corresponds to 65% of the total cytoplasmic phosphatides. There is apparently no qualitative difference between the phosphatides in the microsomes and those in the remainder of the cytoplasm (LEVINE and CHARGAFF 1952).

The earlier view, according to which microsomes have a large complement of oxidative enzymes (cf. MONNÉ 1948) was found to be unjustified after the introduction of better fractionation technique. At present few enzymes are known to be located in the microsomes and still fewer appear to be specifically concentrated there. There seems to be no doubt of the presence of the DPN-cytochrome *c* reductase (HOGEBOOM 1949, HOGEBOOM and SCHNEIDER 1950 c) and the TPN-cytochrome *c* reductase (HOGEBOOM and SCHNEIDER 1950 b), the former being found in higher concentration in the microsomes than in any other fraction. HERS, BERTHET, BERTHET and DE DUVE (1951) have shown that the specific phosphatase glucose-6-phosphatase in liver and in kidney, as well as the alkaline phosphatase of kidney and intestinal mucosa, are located in the microsomes. So marked is this concentration, that glucose-6-phosphatase activity has been employed by DE DUVE, APPELMANS and WATTIAUX (1952) as a test for microsomes. It is probable that further enzymes have yet to be found in the microsomes, as is indicated by the recent studies of SIEKEVITZ (1952), who has shown that the microsomes may be involved in protein metabolism (pp. 83, 109).

B. Mitochondria

The mitochondrial content of the cells varies greatly from tissue to tissue (see also "Morphology"). In rat liver it accounts for about 25%

of the total nitrogen and about 32% of the total protein. The latter value is also obtained by calculation on the basis of succinicoxidase activity. This enzyme, which is totally bound to the mitochondria, affords a suitable index of yield when mitochondria are isolated by centrifugation. For a comprehensive review of analytical data on mitochondria, see Schneider and Hogeboom (1951).

Chemically, mitochondria consist mainly of lipoprotein; in addition they contain small amounts of RNA and of low-molecular components. Proteins, lipids and nucleic acid together account for 92% of the dry weight of the mitochondria (Ada 1949).

1. Nucleic acids

In the estimation of the RNA content of mitochondria, both cytologists and chemists have been confronted by the same difficulty, namely to make observations on the relatively large mitochondria poor in nucleic acid, uninfluenced by the small, but nucleic acid-rich microsomes.

Meyer (1920) was probably the first to show that mitochondria contained nucleic acids. On the basis of the ultraviolet-microscopical studies of Lavin and Pollister (1942) and Harvey and Anderson (1943), Monné (1948) suggests that the mitochondria contain RNA in only insignificant quantities, if at all. On the other hand, Opie and Lavin (1946), also using the ultraviolet absorption technique, have found that RNA is concentrated in large amounts at the periphery of mitochondria in liver cells during hyperplasia induced with dimethylaminoazobenzene (DAB). The autors believe the enhanced basophilic properties of the cytoplasm in hyperplasia to be associated with an increase in the RNA content of the miochondria. This view, strongly supported by Zollinger (1950), is nvertheless contradicted by the findings of Price, Miller, Miller and Weber (1949 a, b), according to which the amount of RNA bound to mitochondria falls off in DAB-induced hepatic tumours, in parallel with the diminution in the mitochondrial population in the tumour tissue.

It appears rather as if the accumulation of RNA at the mitochondrial periphery observed by Opie and Lavin does not occur within the mitochondria, but consists in a perimitochondrial aggregation of microsomes. A counterpart to this phenomenon is to be found in a recent observation by Gustafson (personal communication) according to which the peripheral zone of mitochondria becomes strikingly basophil during the early embryonic development of the sea urchin.

Zollinger (1950), applying the phase-contrast technique to mitochondria incubated with ribonuclease, has obtained information concerning the intramitochondrial distribution of RNA. Such incubation leads to a disappearance of the solid core of the mitochondrion, while the surrounding liquid phase and the membrane are unaffected. Zollinger has thus come to the conclusion that RNA is concentrated in the core. This conclusion finds support in the results of Vendrely (1950), who showed that mitochondria treated with ribonuclease are still stained with iron haematoxylin, while similarly treated microsomes no longer take the stain. Since this dye, according to VENDRELY, stains both RNA and phosphatides, it seems that the microsomes must be completely disintegrated after dissolution of the RNA, while mitochondria, the membranes of which, according to ZOLLINGER, do not contain RNA, can retain their form after the treatment.

ZOLLINGER also comes to the conclusion that the content of RNA in mitochondria is high. In our opinion, however, the total dissolution of the mitochondrial core by ribonuclease does not justify this conclusion: it suffices to assume that RNA occupies a key position in the structure of the core, and that its degradation is accompanied by a dissolution of the latter. Chemical estimation of RNA in mitochondria, a matter forming the subject of a recent exhaustive review by HOGEBOOM and SCHNEIDER (1951), has given progressively lower values as the technique has developed for making mitochondrial preparations free from contamination with microsomes.

CHANTRENNE (1947), and subsequently JEENER (1948), in their studies of the chemical and enzymic heterogeneity of cytoplasmic particles, found that the fraction with the highest sedimentation rate, namely that which represented the purest possible mitochondrial fraction, showed the lowest RNA content. In terms of the dry weight, this content of RNA was about 0.5%.

HOGEBOOM, SCHNEIDER and PALADE (1948) found that the RNA content of mitochondria fell off with the number of washings until it finally attained a constant value. This content, which agreed well with the findings of other workers (SCHNEIDER 1946a, 1948, HOGEBOOM 1949, SCHNEIDER and POTTER 1949, MUNTWYLER, SEIFTER and HARKNESS 1950) had a mean value of 11 μg. RNA-phosphorus per mg. total nitrogen, which corresponds approximately to the percentage value mentioned above. For purposes of comparison, it was recorded that microsomes contained 63 μg. RNA-phosphorus per mg. total nitrogen. This shows that a relatively small contamination with microsomes can bring about a considerable increase in the apparent value for mitochondria. This has apparently occurred in the experiments of several workers, for example BARNUM and HUSEBY (1948).

Mitochondria are Feulgen-negative in sections (OPIE and LAVIN 1946), which indicates that they contain no DNA. ZOLLINGER (1950) has found, however, that after treatment with desoxyribonuclease, mitochondria have a tendency to agglutinate, although the membrane is not dissolved. Small furrows are detected on the membrane, reminiscent of the seams on a football. ZOLLINGER concludes that the mitochondrial membrane contains small amounts of DNA.

2. Lipids

It is agreed by several authors (CLAUDE 1946, SCHNEIDER 1946a, b, BARNUM and HUSEBY 1948, ADA 1949) that the lipid content of mitochondria amounts to 25–30% of the dry weight and that this is mainly composed of phosphatides. LEVINE and CHARGAFF (1952) find that the lipids of mitochondria do not differ in chemical composition from those of the cell as a whole. They nevertheless record the remarkable fact that lipids extracted from isolated mitochondria have different solubilities from those isolated from whole homogenates.

3. Proteins

On the basis of data obtained by Ada (1949), the protein content of mitochondria may be taken as 65—70% of the dry weight. About 60% of the mitochondrial protein is released into solution after disruption of the particles by means of sonic vibrations (Hogeboom and Schneider 1950a). From the purely chemical point of view, the mitochondrial protein has not been the object of detailed study. Schweigert, Guthneck, Price, Miller and Miller (1949) have investigated the amino acid composition and found several deviations from the normal values for liver in tumours induced by DAB. A qualitative study with the aid of the ultracentrifuge has been made by Hogeboom and Schneider (1951), who showed that, of four components detected in normal mouse liver mitochondria, only three could be found in the mitochondria from C3H mouse hepatoma 98/15. Recently de Lamirande, Allard and Cantero (1953) have investigated the variations in the composition of soluble proteins in liver mitochondria during normal and abnormal stimulated growth. Electrophoresis showed an analogous deviation from the normal protein pattern in preneoplastic, neoplastic, and regenerating liver. The authors are inclined to correlate the qualitative changes in the composition of soluble mitochondrial protein with the decrease in the number of mitochondria, accompanying stimulated tissue growth.

4. The enzymic complement of mitochondria

The greater part of the work on the chemical composition of the mitochondria has been devoted to the study of their enzyme complement.

a) Studies on the intracellular distribution of enzymes

Enzymic studies on cytoplasmic particles date from as long ago as the 1910's when Batelli and Stern (1912) and Warburg (1913) observed that the insoluble fraction of cell extracts contained respiratory enzymes. This finding had been well-nigh forgotten when Keilin, in 1929, described the isolation of a particulate enzyme system, the succinic and cytochrome oxidase complex, and was thus the first to isolate an integrated multi-enzyme system.

The first cytological approach to chemical studies of cytoplasmic particles was made in 1934 when Bensley and Hoerr isolated particles from frozen-dried material which they believed to be mitochondria. In the beginning of the nineteen-forties, Claude, Hogeboom and their collaborators (for references, see Claude 1948, Schneider and Hogeboom 1951) worked out a method for the fractionation, by differential centrifugation, of tissues homogenized in isotonic salt solutions, whereby they isolated nuclei, mitochondria and microsomes. This method made it possible to study the distribution of different substances between the constituents of the cell.

Enzymes soon assumed a prominent position in these investigations. In 1946, HOGEBOOM, CLAUDE and HOTCHKISS published their fundamental work on the distribution of cytochrome oxidase and succinic oxidase in the cytoplasm of liver cells. Both enzyme systems were found to be associated with the mitochondrial fraction.

It was already clear at an early stage that the determination of the distribution of an enzyme between the different cell constituents was accompanied by great technical difficulties. The most obvious of these were involved by the necessity of obtaining the fractions in hundred per cent yields and in a complete state of purity. In the case of mitochondria, however, HOGEBOOM, SCHNEIDER and PALADE (1948) have found that, if the preparation is carried out in a hypertonic non-electrolyte medium, the sources of error associated with these difficulties may be minimized. Another improvement in technique was the replacement of the Waring blendor by the Potter-Elvehjem homogenizer.

In order to avoid the erroneous results that may arise from adsorption, SCHNEIDER and HOGEBOOM (1951) have proposed a criterion which must be satisfied if the presence of an enzyme in a given cell fraction is to be regarded as certain, namely that the concentration of the enzyme in the fraction concerned shall be greater than that in the cell as a whole.

Fractional centrifugation in hypertonic (later also isotonic: see p. 7) sucrose solution subsequently became the accepted technique for studies of enzyme distribution which soon acquired great popularity. Data concerning intracellular distribution are now given as a matter of routine in nearly every publication on enzymes. A summary of these data has recently been published by SCHNEIDER and HOGEBOOM (1951).

In spite of this, the information now available concerning typical mitochondrial enzymes is quite scanty. Apart from the aforementioned cytochrome and succinic oxidases, the only enzymes whose presence in the mitochondria can be regarded as certain by the above criterion of SCHNEIDER and HOGEBOOM, are the oxaloacetate and octanoate oxidases, the DPN-cytochrome c reductase and the system which synthetizes p-aminohippuric acid. There is, on the other hand, a long list of enzymes, which are found to be present in isolated mitochondria, without, however, satisfactory data about their relative concentrations in the particles. An enumeration of these enzymes would be of little interest in the present connection.

In considering the data of enzyme distribution, one unfortunately gains the impression that the fractionation procedure, despite the measures taken by the pioneers against the technical difficulties, is still associated with so many insidious pitfalls as to render its value uncertain. Such doubts have been expressed in several quarters (HERS, BERTHET, BERTHET and DE DUVE 1951, HOLTER 1952). This pessimism is all the more justified by the increasing realization that the difficulties are now not so much technical as of a fundamental character.

When it is desired to determine the distribution of an enzyme between a cell constituent and the remainder of the cell, the only possible procedure is to present both fractions with an optimal concentration of substrate. This, however, presupposes that the enzyme in the intact cell actually has access in both fractions to a similar concentration. Our present knowledge of the physiological regulation of enzymic activity indicates that this takes place by a control of the substrate concentrations (Potter and Heidelberger 1950). Many enzymes do not work at full capacity in the living cell (Potter, Recknagel and Hurlbert 1951).

At the same time, it can scarcely be assumed that the concentration of a substrate is uniform throughout the cell. The confinement of an enzyme to a particle surrounded by a membrane, such as a mitochondrion, may be intended to make the access of an enzyme to its substrate easier or more difficult than is the case for another fraction of the same enzyme outside the particle (cf. de Duve, Berthet, Berthet and Appelmans 1951). The fractionation procedure cannot, however, provide information about such matters.

The quantities we measure are hence often inadequate expressions of the activities of enzymes in the cell. Several striking examples of this have recently come to light.

De Duve and his co-workers (de Duve, Berthet, Berthet and Appelmans 1951, Berthet and de Duve 1952, Berthet, Berthet, Appelmans and de Duve 1952, de Duve, Appelmans and Wattiaux 1952) have shown that an acid ester phosphatase found enclosed in a mitochondrial fraction, cannot act upon added β-glycerophosphate unless the particles are first made permeable to the glycerophosphate by treatment with surface-active substances or electrolytes. Glycerophosphate, like sucrose, has a protective action on the mitochondrial structure as a consequence of its low penetrating power.

At about the same time, Kielley and Kielley (1951) reported that mitochondria prepared in a sucrose medium showed almost no ATPase activity, although such activity could be developed if the particles were subjected to ageing, hypotonic media, or other treatment causing structural disintegration. This finding, subsequently confirmed by Potter and Recknagel (1951), has not yet been fully explained. So much, however, appears to be clear, that ATPase as such is present in a non-operative state in the intact mitochondrion. There is now reason to suspect that those enzymes which appear as ATPases after a structural disintegration, are actually fragments of phosphate-transferring enzymes. This matter is discussed in further detail on p. 69.

The importance of these examples as warnings against premature conclusions from enzyme distribution data, is shown by the fact that as recently as in 1951, Schneider and Hogeboom, with the support of several data from the literature, stated that the acid phosphatase and ATPase were concentrated in the mitochondria. It is possible that reports concerning the distribution of the adenylic acid phosphatase are equally premature. Novikoff, Podber and Ryan found in 1950 that this enzyme was concentrated

in the mitochondria. In the following year KIELLEY and KIELLEY reported
that their preparations contained no trace of the enzyme.

The above examples also serve to emphasize the difficulty of establishing, from the intracellular distribution of an enzyme, the part played
by a given cell constituent in the metabolic function associated with that
enzyme. In these cases and, indeed, in all cases where phosphatases are
concerned, we have very little idea of their physiological rôle. In other
instances the position is perhaps clearer, but there remains the fundamental
question of whether an enzyme, found in several locations within the cell,
has the same function at all these locations.

The enzyme glucokinase, to which we shall often have occasion to
refer, proves, in fractional centrifugation experiments, to be largely bound
to particles (CRANE and SOLS 1953). Let us imagine that we have a method
at our disposal whereby we can isolate cell membranes from the remainder
of the cells. We should then find that the greater part of the glucokinase
was in the residual fraction. In other words, we should easily gain the
impression that the cell surface played only a minor part in glucokinase
activity. We know, however, that the surface layer of the cell performs
an active transport of glucose by a glucokinase mechanism. The quantitatively insignificant amount of glucokinase in the surface layer is thus
responsible for a very important specific function. The frequently mooted
example of the distribution of isocitric dehydrogenase between the mitochondria and the ground cytoplasm perhaps comes, from the critical
standpoint, under the same heading. This question will be discussed fully
at a later stage.

It is in no way intended by the above remarks to detract from the
important part played by enzyme distribution studies by means of
fractional centrifugation in the development of our present concept of
the physiological function of the mitochondria. In our opinion, however,
the value of the method is most obvious when it furnishes evidence for
the *exclusive* location of an enzyme in a certain cell constituent. This
type of information serves as a useful pointer in the search for the metabolic function of the cell constituent concerned. For the detailed
investigation, however, studies of multi-enzyme systems seem to be more
suited. Thus, the finding that cytochrome oxidase and succinic oxidase are
located exclusively in the mitochondria, has been of the greatest importance in that it directed attention upon the central rôle of the mitochondria in cell respiration. The systematic investigation, however, became
the task of the enzymologist, when GREEN and his group at Wisconsin
University, inspired by the tradition of KEILIN, commenced their studies
of the cyclophorase system.

b) The cyclophorase concept

In 1947, GREEN, LOOMIS, DICKMAN, AUERBACH and NOYCE reported that
a particulate preparation from rabbit kidney catalyzed the oxidation of
Krebs cycle intermediates with the accumulation of inorganic pyrophosphate. The first systematic study of this system was published in the

following year, when Green, Loomis and Auerbach (1948) reported that washed particles from tissues homogenized with a Waring blendor in isotonic salt solution, contained the entire complement of enzymes and coenzymes necessary for the total oxidation to carbon dioxide and water of pyruvate and all the intermediates of the Krebs cycle. The enzyme preparation and the enzyme system contained therein were named cyclophorase and the cyclophorase system respectively, in order to emphasize that the system was more than a simple mixture of enzymes and coenzymes.

In the succeeding years, the Green group and a number of other workers devoted themselves to a painstaking investigation of this enzyme system (cf. Green 1951 b). From these studies—publications concerning the cyclophorase system are now over thirty in number—there was gradually evolved a picture of a multienzyme system very different from a random mixture of its constituent enzymes. The essential differences may be summarized as follows:

1. The ratios in which the enzymes are present are such that the oxidation of Krebs cycle intermediates to carbon dioxide and water proceeds without the accumulation of intermediates.

2. The system includes the coenzymes essential to its function. These ensure the full catalytic capacity, despite that they are present in much lower concentrations than are necessary to give a corresponding activity in an unorganized enzyme mixture. If, however, the organization is destroyed by subjecting the system to ageing, hypotonic conditions or other treatment, a considerably higher concentration of coenzymes is necessary for the restoration of the catalytic power.

3. The energy liberated in the oxidation is fixed in the phosphate-bound form.

These phenomena led Green to regard the cyclophorase system as a macromolecular conjugate of the constituent enzymes and the coenzymes indissociably bound to them.

Green found support for the cyclophorase theory in the observation that the particles and the surrounding medium were not in equilibrium with respect to concentrations of coenzymes and related substances. Thus Teply (cf. Green 1951 b) found in the particles relatively high concentrations of those vitamins which are known to be constituents of coenzymes (Table 2). Huennekens and Green (1950) showed that added DPN underwent cleavage in the presence of the particles, while the "conjugated" coenzyme was not attacked.

Green, Atchley, Nordmann and Teply found in 1949 that cyclophorase preparations, after repeated washings with a phosphate-free medium, still contained inorganic orthophosphate, the quantity of which could be increased by aerobic incubation in the presence of orthophosphate, irrespective of the concentration of the latter. This phosphate, called by the authors gel phosphate, was considered to be a hydrolysis product of a extremely labile phosphate ester conjugated with the enzyme complex. Holzer and Lynen (1950) have found a substance with analogous properties in yeast. Albaum (1949) has analysed cyclophorase for adenosine polyphosphates and found significant quantities of ADP and ATP, together with a nucleotide containing labile phosphate but not identical with either of these.

While the Wisconsin group were engaged in their detailed study of the cyclophorase system and developing the theory outlined above, SCHNEIDER and POTTER (1949) and KENNEDY and LEHNINGER (1948, 1949) found, independently of one another, that isolated mitochondria could catalyse the oxidation of a number of Krebs cycle intermediates and that of fatty acids, and could effect the conversion of the oxidative energy to the phosphate-bound form.

These observations led GREEN (1951 a, b) to identify cyclophorase with the mitochondria and to suggest that the latter represent the structural units of the cyclophorase complex. GREEN found evidence in support of this hypothesis in the observations of HARMAN (1950 a), who established a correlation between the form and number of the mitochondria and the cyclophorase activity in a homogenate.

Table 2. *Content of some vitamins in cyclophorase.*
After TEPLY (quoted from GREEN 1951 b).

Vitamins	Liver	Kidney
	μg/g dry weight	μg/g dry weight
Niacin[1]	330	275
Riboflavin[1]	120	110
Biotin	0.7	0.9
B_6	33	22
Pantothenic acid[1] .	225	175
Folic acid	50	40

[1]) Present in the form of corresponding coenzymes.

c) The relation of cyclophorase to mitochondria

SCHNEIDER and HOGEBOOM (1951) have seriously contested this identification of the cyclophorase system with the mitochondria. They allow that the oxidation of α-ketoglutarate and oxidative phosphorylation may be considered to be associated with the mitochondria, but they believe evidence to be lacking that pyruvate can undergo complete oxidation to carbon dioxide and water in these particles, which hence, by definition, cannot be identical with cyclophorase. They cite a publication of KENNEDY and LEHNINGER (1948), who showed that the oxidation of pyruvate and malate resulted in a partial accumulation of citrate. The experiment was made with mitochondria prepared in sucrose. It may, however, be questioned whether this is a suitable example, since the mitochondria were limited in their oxidative activity by a lack of phosphate acceptors. The latter were present only in catalytic amounts, while no trapping system for newly formed ATP was added. It was not known at the time of this investigation that sucrose mitochondria lack ATPase activity and the power of accumulating inorganic pyrophosphate (KIELLEY and KIELLEY 1951). More recent experiments (BRODY and BAIN 1952, PRESSMAN and LARDY 1952) made under proper experimental conditions with mitochondria prepared in sucrose, show that pyruvate can be oxidized as readily as α-ketoglutarate.

A further objection raised by SCHNEIDER and HOGEBOOM against GREEN's hypothesis is that, in their opinion, the presence of isocitric dehydrogenase in mitochondria cannot be regarded as proved. According to the enzyme distribution data, only about 12% of the total quantity of this enzyme in the cell is bound to the mitochondria (Hogeboom and Schneider 1950). Cyclophorase preparations oxidize citrate as vigorously as other metabolites of the Krebs cycle (Green, Loomis and Auerbach 1948, Paul, Fuld and Sperling 1953). Sucrose mitochondria, on the other hand, have been found by several workers (Plaut and Plaut 1952, Lindberg, Ljunggren, Ernster and Révész 1953) to attack added citrate only very weakly or not at all. In view of the low percentage of the isocitric dehydrogenase found in the mitochondria, authors other than Schneider and Hogeboom have also been inclined to associate this with the failure of citrate to be oxidized. The matter has recently been investigated thoroughly by Plaut and Plaut (1952). With the aid of isotopes, they have found that the phenomenon is not due to a low isocitric dehydrogenase activity in the particles. In view of the results of inhibition experiments with fluorocitrate, recently presented by Peters (1952), and of Plaut's work which is described more fully at a later stage, it appears reasonable to attribute the inertness of mitochondria towards citrate to a permeability barrier.

This is a further illustration of the danger involved in drawing conclusions from enzyme distribution data. If there is a possibility of a permeability barrier towards a given substrate, a worker seeking reliable information of the distribution of the corresponding enzyme is confronted by a dilemma: shall an effort be made to keep the mitochondrial membrane as intact as possible in order to avoid leakage of the enzyme, or shall the membrane be destroyed so that the substrate may have access to the enzyme?

Even if we disregard this difficulty and accept the distribution ratio of 12/88 for the isocitric dehydrogenase, we must take into consideration the point previously emphasized, namely that the importance of an enzyme in a given cell constituent can hardly be judged in relation to the amount of the same enzyme found elsewhere in the cell. In the present case, the fact that the isocitric dehydrogenase is bound to the mitochondria in only a small fraction of its total amount, by no means implies that this small fraction cannot carry out its function with perfect efficiency in a reaction chain, without even constituting a limiting factor (cf. Potter, Recknagel and Hurlbert 1951).

We cannot, by and large, see any reason why Green's hypothesis, according to which the cyclophorase system is associated with the mitochondria, should not be accepted. A wholehearted acceptance of the theory is nevertheless rendered somewhat difficult by the biased attitude assumed by the Green school towards recent discoveries concerning the mitochondrial structure, and especially towards the question of the mitochondrial membrane. This attitude has been sharply criticized in several quarters (Schneider and Hogeboom 1951, Hers, Berthet, Berthet and de Duve 1951).

The cyclophorase concept has been developed on the basis of experiments with particles prepared in a salt medium from homogenates made in the Waring blendor. Such preparations, according to Green (1951 b) have a gel-like consistency and lack selective permeability. They nevertheless show osmotic properties, swelling in hypotonic and shrinking in hypertonic media. In order to explain the fact that they contain a complement of low-molecular components, especially coenzymes, it would therefore seem reasonable to assume the presence of a semipermeable membrane. This is also suggested by the finding of LEHNINGER (1951) that "externally" supplied $DPNH_2$ is not capable of giving rise to phosphorylation in cyclophorase preparations unless the particles are treated with

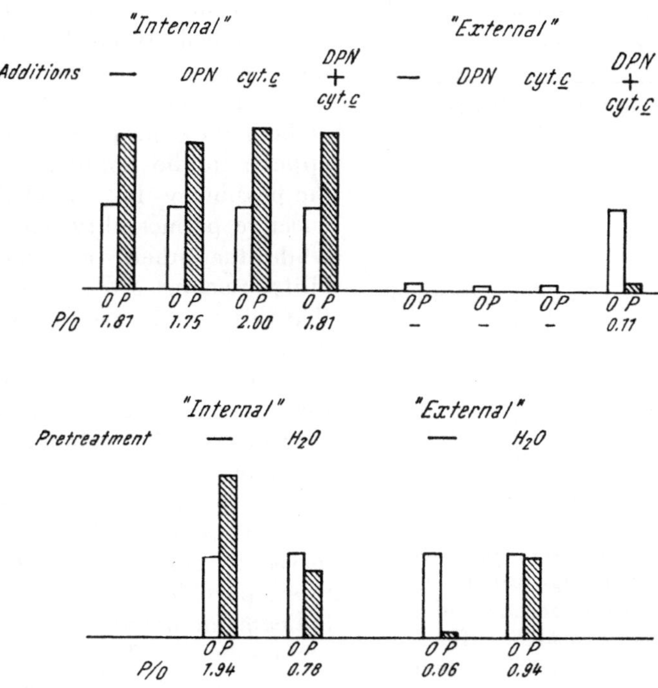

Fig. 7. Phosphorylation coupled to the oxidation of $DPNH_2$ in liver mitochondria. *"Internal"*: $DPNH_2$ generated *within* the mitochondria, by oxidation of β-hydroxybutyrate. *"External"*: $DPNH_2$ generated *outside* the mitochondria, by oxidation of ethanol by means of added alcohol dehydrogenase. The system requires external addition of DPN and cytochrome c for oxidation, and pretreatment of the mitochondria in hypotonic medium is needed for coupled phosphorylation. (From LEHNINGER 1951.)

distilled water for about 20 minutes. Moreover, the cytochrome c associated with the particles cannot function as electron transmitter in the oxidation of "external" $DPNH_2$ (Fig. 7).

In the opinion of GREEN (1952) and HARMAN (1950 a) these phenomena are equally compatible with a protein gel structure which retains the coenzymes in an undissociable state.

As criteria for the existence of a membrane, these authors require that the particles should show a tendency to lysis and display selective permeability. With regard to the former point, HARMAN's finding that lysis does not occur in cyclophorase preparations stands in direct contradiction to the results of other workers (cf. SCHNEIDER and HOGEBOOM 1951).

Before we consider the second requirement, let us first consider to what extent selective permeability is a suitable criterion for the existence of a membrane. As far as we can judge, semi-permeability and selective permeability are independent concepts: the first is a passive property, the second active, probably enzymic, in nature. If, therefore. GREEN and HARMAN dispute the presence of a semipermeable membrane, their criterion of selective permeability is unjustified. If, on the other hand, they consider that the assumption of a membrane implies its possession of both these properties, then there appears to be nothing to exclude the possibility that one of them, the selective permeability, may be lost, while the other, the semipermeability, remains. We know, in fact, that this must be the case with dried yeast cells.

Table 3. *Permeability data on rabbit heart sarcosomes.*

(After CLELAND 1952.)

Substance tested	Permeability
Sucrose	0.006
NaCl	0.013
KCl	
K-succinate	
K-fumarate	0.02 (approx).
K-malate	
$H_2PO_4^-$	
Glucose	0.024
Glycerol	
Ethyleneglykol	> 10 [1])
HPO_4^{--}	

[1]) not measurable.

The sarcosomes were prepared in a medium containing 0.01 M versene (ethylene diamine tetraacetic acid). Permeability was measured, after suspension of the mitochondria in buffered saline containing versene, by registering light scattering in a Beckman spectrophotometer at 520 mμ. Permeability is expressed as the reciprocal of the time (in minutes) required for a change of 5 per cent in the initial density of the suspension. Varying pH or ATP concentration alter the values.

Mitochondria prepared in a sucrose medium do not show the gel consistency of cyclophorase preparations (SCHNEIDER and HOGEBOOM 1951). In addition to the simple osmotic properties which can be observed in cyclophorase, these more intact particles have qualities which indicate an active function in the surface layer. The previously cited experiments of DE DUVE and his coworkers (DE DUVE, BERTHET, BERTHET and APPELMANS 1951, BERTHET and DE DUVE 1952, BERTHET, BERTHET, APPELMANS and DE DUVE 1952) suggest that these mitochondria possess a surface structure which can, in some degree, influence enzymic reactions, separate enzymes from their substrates and selectively control the passage of substances. The latent ATPase of KIELLEY and KIELLEY (1951) provides even stronger evidence that the surface structure plays a metabolic rôle. It also follows from the experiments of these authors that intact liver mitochondria, in contrast to earlier findings with liver cyclophorase (CROSS, TAGGART, Covo and GREEN 1949), cannot accumulate inorganic pyrophosphate during respiration, so that their oxidative activity may be controlled by addition of hexokinase. The surface of an intact mitochondrion may hence have an active part in mediating an external control of the cyclophorase activity.

A further stage in the demonstration of selective permeability in the intact mitochondrion was reached with the recent studies by CLELAND (1952) of heart mitochondria prepared in a medium containing versene. This compound binds calcium ions in complex linkage. CLELAND found that these particles showed a marked selective permeability. Using a light scattering technique—a method previously employed in the study of the permeability of erythrocytes—he obtained quantitative values for the relative penetration rates of a series of cations, anions and non-electrolytes. His results are reproduced in the accompanying table (Table 3).

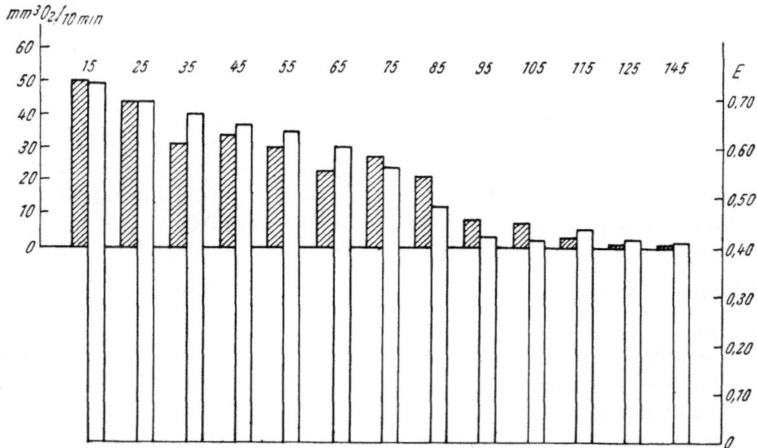

Fig. 8. Decrease of O₂-consumption in mitochondria swollen after incubation without substrate.
Each of the 13 Warburg vessels contained 0.5 ml mitochondrial suspension ; 0.3 ml ATP, 0.01 M ; 0.1 ml MgSO₄, 0.1 M ; 0.2 ml isotonic K-phosphate buffer, pH 7.2 ; 0.6 ml mannitol, 5.75 %. Side arm : 0.3 ml Na-pyruvate, 0.08 M, tipped in after different intervals of time (as indicated, in minutes, at the head of each double column). Black figures : O₂-consumption after introduction of pyruvate. White figures : extinction (E) of the sample, diluted at the end of the experiment, with 20 volumes of 5.75 % mannitol. The extinction value is inversely proportional to the degree of swelling of the mitochondria.
(From RAAFLAUB 1952)

The figures show that the selective permeability especially concerns electrolytes which, with the exception of secondary phosphate, can pass the mitochondrial membrane only with difficulty. The penetration of hydrophilic non-electrolytes. such as sucrose, glucose and glycerol, seems to be determined mainly by the molecular size. ATP in a concentration of 0.0025 M has an effect resembling that of versene, in that it reduces the permeability of the mitochondria to electrolytes. RAAFLAUB (1952) has observed that swelling tendes to depress the respiration rate of mitochondria (Fig. 8). He found that ATP prevents the swelling of the particles. It is interesting to recall that a protective effect of ATP has been observed on the cell surface of sea urchin eggs and amoebae (KRISZAT 1949, 1950, 1951, 1952, RUNNSTRÖM and KRISZAT 1950 a, b. KRISZAT and RUNNSTRÖM 1951, 1952).

It may also be of interest to record preliminary values obtained by CLELAND for the chemical constitution of versene mitochondria, since these are. as far as we are aware, the first analytical results for a mitochondrial material in which the leakage of substances through the damaged membrane has been consciously avoided (Table 4).

These experimental findings incline us to visualize a gradient of surface organization from the cyclophorase preparations of Green to the selectively permeable mitochondria, prepared in versene, at the highest level. At the cyclophorase level, this surface organization is responsible for a more or less intact osmotic system of a passive character; at the level of the intact mitochondrion, there is also an active enzymic organization which makes possible selective permeability and a metabolic regulation. On this basis, the cyclophorase system will represent a level of organization derived from the mitochondrion, just as the organization of Keilin's enzyme complex is derived from that of the cyclophorase system.

Table 4. *Chemical constitution of cat heart sarcosomes prepared in sucrose medium containing versene.*

(Unpublished analytical values by Cleland.)

Lipids

Total	0.38 mg/mg protein
Phospholipids	0.31 „ „
Sterols	0.004 „ „

Ribonucleic acid .... 0.007 „ „

Anions

Chloride	< 0.02 μM/mg protein
Phosphate, inorganic	0.064 „ „
Phosphate, acid-labile	0.036 „ „
Phosphate, total acid-soluble	0.139 „ „

Cations

Ca^{++}	0.099 μM/mg protein
Mg^{++}	0.29 „ „
Na^+	0.05 [1] „ „
K^+	0.15 „ „

[1] approximately.

We therefore consider that, while the cyclophorase system is an integral part of the mitochondrion, this is no justification for regarding the two systems as equivalent and the designations freely interchangeable. This has actually been done in several instances and has been rightly criticized with some vigour by Schneider and Hogeboom (1951). There has recently been a tendency on the part of the Wisconsin group to distinguish between the two terms by giving the name "cyclophorase" to particles prepared in salt media and reserving the term "mitochondria" for preparations made in sucrose (Harman and Feigelson 1952 a, c).

When we deal later in some detail with the enzyme complement of mitochondria, using the cyclophorase system as our starting point, we do not propose to make a distinction of this kind, since most of the information gained from cyclophorase applies also to mitochondria. In those matters, however, where the differences between cyclophorase and mitochondria are directly concerned, such as the case of ATPase, the aforementioned distinction will be made.

Enzymic Processes Concerned in the Physiological Function of Mitochondria

In a systematic treatment of the enzymic processes related to the physiological function of the mitochondria, a division under the following headings appears to be suitable:

I. E n e r g y - g e n e r a t i n g p r o c e s s e s, including a review of the oxidative processes associated with the mitochondria, theoretical and experimental values for the energy yield in these processes and the mechanism whereby this energy is preserved in the form of the so-called energy-rich bonds.

II. E n e r g y - t r a n s f e r r i n g p r o c e s s e s, where a description is given of the processes in the mitochondria where energy in the form of energy-rich bonds is transferred reversibly to the adenylic acid system.

III. E n e r g y - u t i l i z i n g p r o c e s s e s, covering an account of synthetic processes in the mitochondria and of the relation of other synthetic processes to the energy-generating and energy-utilizing functions of the mitochondria.

Before we enter into a detailed treatment of these matters, it may be well to consider the general import of the concept of the energy-rich bond.

Energy-rich bonds are taken, in this connection, to include a number of types of chemical linkage, of biological importance, with a value of free energy of hydrolysis $(-\triangle Fo)$ in the region of 10,000 cal. In the original concept, for which KALCKAR (1941) and LIPMANN (1941) were responsible, the inclusion of phosphate groups was a specific feature of these bonds. The bonds said to be of biological importance were:

$$
\begin{array}{ll}
\overset{\displaystyle O}{\overset{\|}{C}} \quad O \sim P(OH)_2 . O & \text{(as in acetylphosphate)} \\[2ex]
\overset{\displaystyle C}{\overset{\|}{C}} \quad O \sim P(OH)_2 . O & \text{(as in phosphopyruvate)} \\[1ex]
= N \sim P(OH)_2 . O & \text{(as in phosphocreatine)} \\[1ex]
\overset{O}{P} \quad O \sim \overset{\overset{\displaystyle O}{\|}}{P} & \text{(as in adenosinetriphosphate)} \\[1ex]
OH \qquad OH &
\end{array}
$$

Recently, by the work of LYNEN and his co-workers (LYNEN and REICHERT 1951, LYNEN, REICHERT and RUEFF 1951, HOLZER 1952), it has been made clear that other types of energy-rich bonds, such containing sulphhydryl groups, are also of great biological significance. This class includes

$$
\begin{array}{ll}
S \sim \overset{\overset{\displaystyle O}{\|}}{C} & \text{(as in acetyl-CoA), and possibly} \\[1ex]
S \sim P(OH)_2 . O & \text{(as in phosphoryl-CoA)}
\end{array}
$$

As will be discussed in the following account, energy-rich bonds of the sulphhydryl type are those formed primarily in connection with oxidative processes [1].

The energy from these primary carriers can be transferred reversibly to a number of intermediates of both the sulphhydryl and the phosphoryl types and is thus made available in various specific forms for the different synthetic processes (e. g. as acetyl-CoA for acetylation processes, as ATP for phosphorylation of glucose, etc.).

I. Energy-Generating Processes

A. Mitochondrial Oxidations

1. The Krebs cycle

In the present conception of the Krebs cycle, a molecule of "active" (CoA-bound) acetate is condensed with a molecule of oxaloacetate with the formation of citrate (for the mechanism of the condensation, cf. p. 74). By a series of oxidations, decarboxylations and other transformations, this citrate is broken down to oxaloacetate which can then condense with a further molecule of "active" acetate. The individual reactions participating in this cyclic process have, in a large degree, been finally established and are shown in Fig. 9. (For recent findings concerning the aconitase reaction, cf. MARTIUS and LYNEN 1950 and MARTIUS 1952.) The net result of a revolution of this cycle is that a molecule of acetate is oxidized to carbon dioxide and water. GREEN and his school (for review see GREEN 1951 a, b) have made a fundamental contribution to our knowledge by showing that the entire enzyme complement which catalyses the reactions of the Krebs cycle is located in the mitochondria.

The substance most closely related metabolically to the cycle is pyruvate. This can be oxidatively decarboxylated in the mitochondria to "active" acetate and is thus also oxidized to carbon dioxide and water. Malate and oxaloacetate constitute, so to speak, a junction point in the cycle. They can both, by means of two specific enzymes, OCHOA's (1951) "malic enzyme" and the oxaloacetate carboxylase respectively, be converted into pyruvate (Fig. 10). A consideration of the mechanism of these reactions follows on p. 78. As these enzymes are also present in the mitochondria, all the intermediates of the Krebs cycle can be oxidized there to carbon dioxide and water. The oxidation of oxaloacetate, however, requires an equivalent amount of "active" acetate and the presence of ATP.

---

[1] Energy-rich bonds may also be formed *non-oxidatively*, in the thioclastic cleavage of β-ketoacids and in the removal of water from a phosphorylated secondary alcohol (enolisation). The former case will be dealt with subsequently (p. 36); the latter, of which the only known example is the reaction:

$$\text{2-phosphoglyceric acid} \xrightarrow[- H_2O]{\text{enolase}} \text{phospho}enol\text{pyruvic acid}$$

does not appear to occur in mitochondria.

The function of the latter in this connection is not yet clear (GREEN, BEINERT, FULD, GOLDMAN, PAUL and SARKAR 1953).

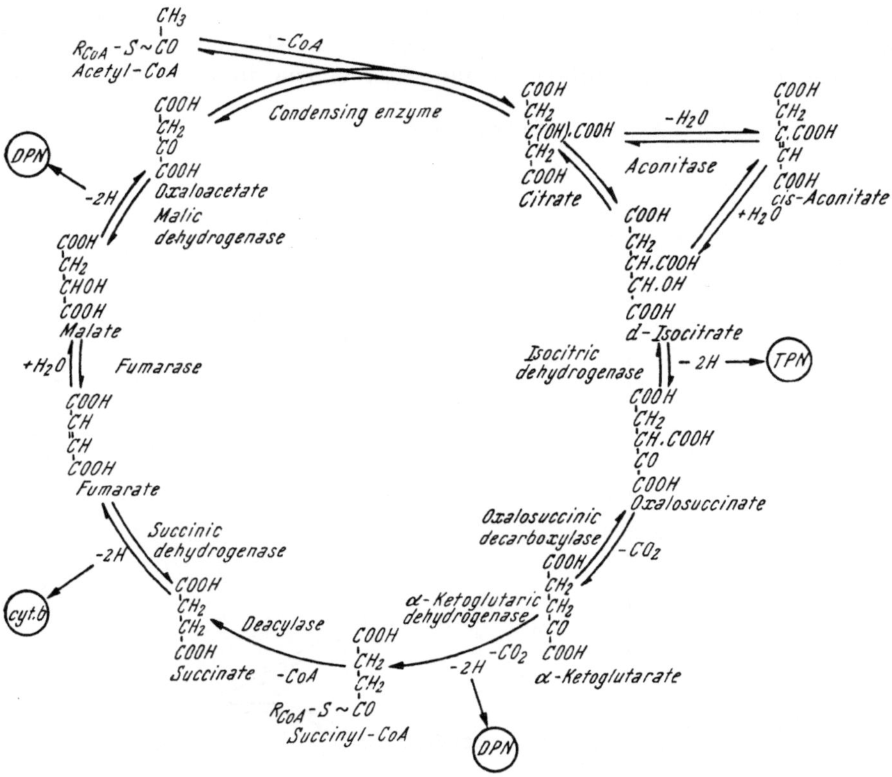

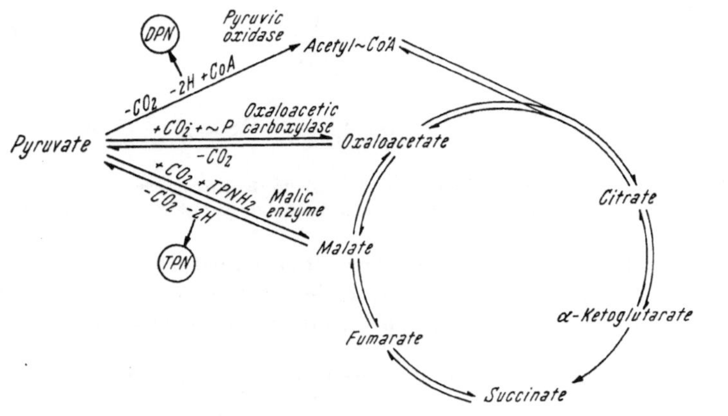

Fig. 9. Oxidation of "active" acetate by way of the Krebs cycle.

Fig. 10. The relation of pyruvate to the Krebs cycle.

2. Oxidations related to the Krebs cycle

The number of substrates which can undergo complete degradation in the mitochondria is not, however, hereby exhausted. In general it may be said that the condition for total degradation is that the substrate in question shall be capable of direct or indirect transformation by means of enzymes and co-factors present in the mitochondria, into one of the inter-

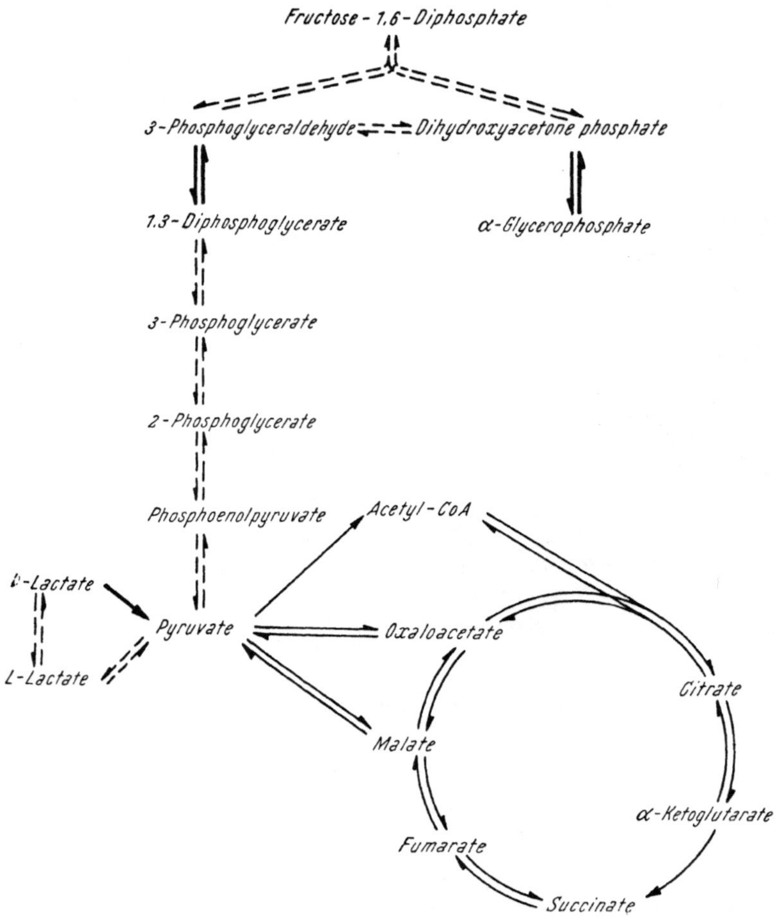

Fig. 11. The relation of glycolysis to the Krebs cycle.
The reactions marked by dotted arrows are not known to be catalyzed by mitochondria.

mediates of the cycle. If, however, this transformation gives rise to "active" acetate or pyruvate, a catalytic concentration of another intermediate of the cycle is necessary to initiate the condensation to citrate.

a) The oxidative reactions in glycolysis

It has recently been reported by Kaplan, Still and Mahler (1951) that the oxidative reactions of glycolysis (Fig. 11), *i.e.* the oxidations of

3-phosphoglyceraldehyde and α-glycerophosphate, have been found to be catalyzed by isolated particles.

Lactate is oxidized in the mitochondria to pyruvate by an enzyme which, in contrast to the long-known lactic dehydrogenase of the ground cytoplasm, is d-specific (MAHLER, TOMISEK and HUENNEKENS 1953). If l-lactate is to be oxidized with the aid of the mitochondrial enzyme, the mitochondria must be supplemented with a cytoplasmic enzyme, the α-hydroxy acid racemase, which racemices optically active modifications of different α-hydroxy acids (HUENNEKENS, MAHLER and NORDMANN 1951).

b) Fatty acids

The oxidation of fatty acids in the mitochondria has in recent years been the object of detailed study by LEHNINGER and by GREEN and their associates (for review, see GREEN 1951 b) (Fig. 12). The condition for the oxidation of a fatty acid in the mitochondria is the presence of a dicarboxylic acid intermediate of the Krebs cycle, a so-called "sparker". The fatty acids are oxidized in accordance with the classical β-oxidation principle of Knoop (for review, see BREUSCH 1952) and are hence presented

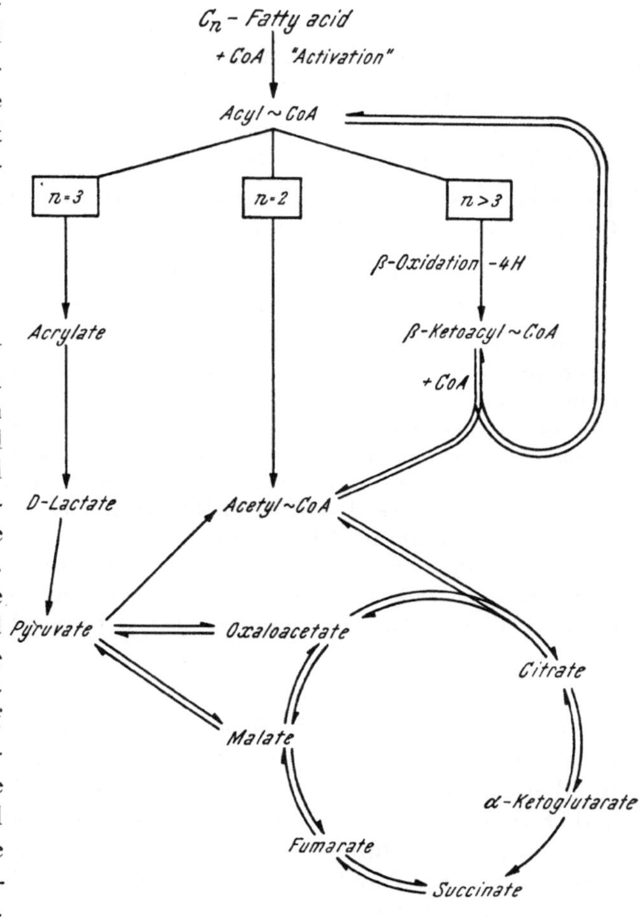

Fig. 12. The relation of the oxidation of fatty acids to the Krebs cycle.

to the Krebs cycle in the form of C_2 units. The sparker, however, has not the exclusive function of condensing partner, but also serves to provide the initial energy required for the "activation" of the fatty acids before they can undergo β-oxidation. The nature of this initial activation has been revealed in recent years, largely by findings concerning the rôle of coenzyme A in the fatty acid metabolism of microorganisms (for review, see BARKER 1951). It consists in the coupling of the fatty acid with CoA by a reversible reaction, the detailed mechanism of which is still unclear. According to Green (1952), the reaction proceeds in the following manner:

$$\text{phosphoryl-CoA} + \text{fatty acid} \rightleftharpoons \text{acyl-CoA} + \text{phosphate}$$

The "active" fatty acid, the formula of which may be written $R_{CoA} - S \sim CO \cdot (CH_2)_n \cdot CH_3$, is oxidized in two stages by β-oxidation:

$$R_{CoA} - S \sim CO \cdot (CH_2)_n \cdot CH_3$$

$$\downarrow -2\,H$$

$$R_{CoA} - S \sim CO \cdot CH = CH \cdot (CH_2)_{n-2} \cdot CH_3 \;\; \underset{-H_2O}{\overset{+H_2O}{\rightleftharpoons}} \;\; R_{CoA} - S \sim CO \cdot CH_2 \cdot CHOH \cdot (CH_2)_{n-2} \cdot C$$

$$\downarrow -2\,H$$

$$R_{CoA} - S \sim CO \cdot CH_2 \cdot CO \cdot (CH_2)_{n-2} \cdot CH_3$$

The β-keto compound thus formed is split with the aid of CoA with the formation of acetyl-CoA and a shortened "active" fatty acid:

$$R_{CoA} - SH + R_{CoA} - S \sim CO \cdot CH_2 \cdot CO \, (CH_2)_{n-2} \cdot CH_3 \rightleftharpoons$$
$$\rightleftharpoons R_{CoA} - S \sim CO \cdot CH_3 + R_{CoA} - S \sim CO \cdot (CH_2)_{n-2} \cdot CH_3$$

The oxidations of C_3 and C_4 fatty acids constitute special cases. When, in accordance with the above scheme, the oxidation has proceeded to acetoacetyl-CoA or to propionyl-CoA, according to whether the initial fatty acid molecule had an even or an odd number of carbon atoms, the degradation continues in a manner dependent upon the nature of the organ. In the liver the oxidation ceases at the acetoacetate stage, while propionate is oxidized to pyruvate by way of acrylate and d-lactate (Mahler, Tomisek and Huennekens 1953). In kidney and muscle the circumstances are reversed (cf. p. 66).

The oxidation of unsaturated and branched fatty acids and of certain fatty acid derivatives such as amines can also proceed to a certain extent in the mitochondria, although the mechanisms are not at present known.

c) Amino acids

Of the amino acids, glutamate occupies a special position, in that it can be oxidized to carbon dioxide and water after oxidative deamination to a-ketoglutaric acid, without the presence of other metabolites (Taggart and Krakaur 1949, Still, Buell and Green 1950 a). For the oxidative deamination of alanine (Still, Buell and Green (1950 b) and aspartate (Still, Buell, Knox and Green 1949, Nakada and Weinhouse 1950) catalytic quantities of a-ketoglutarate are required (cf. Fig. 13).

Phenylalanine and tyrosine are also transaminated in the presence of a-ketoglutarate to phenylpyruvate and p-hydroxyphenylpyruvate, respectively. The transaminases for these latter amino acids have been found to be exclusively located in the mitochondria (Hird and Rowsell 1950). It is not yet known, however, how the resulting keto acids are coupled with the Krebs cycle. A further transamination reaction is the recently discovered conversion of cysteic acid to thiolpyruvic acid in the presence of a-ketoglutarate (Darling 1952). It has not yet been investigated whether this reaction is confined to the mitochondria.

Tryptophane, valine, leucine, histidine and glycine are deaminated in the mitochondria in the presence of α-ketoglutarate (HIRD and ROWSELL 1950). With the exception of glycine, it seems probable that these acids are converted to the corresponding keto acids, the fates of which, as in the case of the above-mentioned phenylpyruvate, are unknown. Nevertheless, the degradation of all these amino acids commences in the mitochondria.

It is therefore reasonable to suppose that the keto acids formed are in some way converted into compounds—fatty acids or intermediates of the Krebs cycle—which can finally undergo complete degradation in the mito-chondria. It is hence tempting to conclude that the intermediate reactions are also mitochondrial processes, provided that the necessary co-factors and concomitant reactions are present.

The same considerations apply to the oxidation of glycine. It has now been shown that glycine can enter two metabolic pathways, one consisting in the oxidative removal of a C_1 body and the other in a condensation with

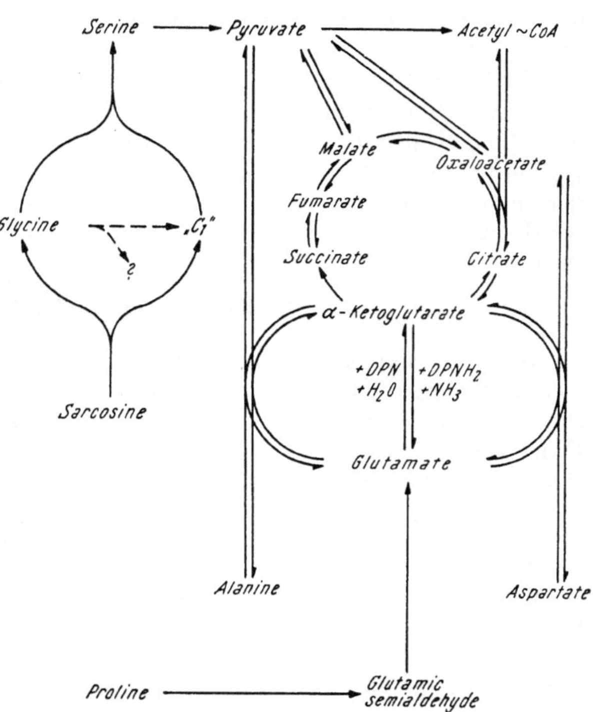

Fig. 13.   The relation of the oxidation of some amino acids to the Krebs cycle.

an "active" C_1 body to form serine which in turn is oxidatively deaminated to pyruvate (SAKAMI 1948, SIEKEVITZ and GREENBERG 1949). Of these processes, the oxidative deamination of serine is known to be located in the mitochondria (GREEN 1951 b), as also is the splitting of glycine, which actually follows from the above-mentioned experiment, where glycine was shown to yield its amino group to α-ketoglutarate (HIRD and ROWSELL 1950). The metabolic cycle cannot be completed, since the condensation between the "active" C_1 body and glycine has not yet been elucidated. In view of the fact that an "active" C_1 compound is said to be concerned here, it is conceivable that a solution of the matter is to be sought among the reactions and possible co-factors that may be involved in the "activation" of the C_1 body. A possible association of this mechanism with the mitochondria is to be found in the fact that the co-factors which are believed to be concerned, such as folic acid (TEPLY, cf. GREEN 1951). folinic acid and vitamin B_{12}

(Swendseid, Bethell and Ackermann 1951), are found in quite high concentrations in the mitochondria.

It has recently been reported that a number of processes involving a degradation of glycine derivatives have also been observed to occur in the mitochondria (Green 1951 b). These include the oxidative demethylation of sarcosine to glycine and formaldehyde (Handler, Bernheim and Klein 1941), the decarbamylation of guanidoacetic acid in the presence of ornithine with the formation of glycine and arginine, and the degradation of glutathione.

Finally, it has been possible to locate the breakdown of proline and hydroxyproline in the mitochondria (Taggart and Krakaur 1949). The oxidation of proline has been shown to take place in two stages, with glutamine-δ-semialdehyde as the probable intermediate. This compound may give rise to speculation as to the existence of an "active" glutamate, which may conceivably act in the synthesis of glutamic acid derivatives such as glutamine.

3. Miscellaneous oxidations

Another group of oxidations found to take place in the mitochondria comprises single oxidative attacks on diverse substrates unconnected with the citric acid cycle. These include the oxidation of choline to betaine (Williams 1951, 1952 a, b) and that of ascorbic acid to dehydroascorbic acid (Friedkin and Lehninger 1949 b).

4. The electron transfer system

Most of the substrates dealt with in the preceding sections are oxidized by pyridine nucleotides, whence the electron transport proceeds by way of flavin nucleotides and the cytochrome system to oxygen. An exception is found in the case of succinate, which is directly oxidized by the cytochrome system. Some substrates, such as xanthine, are coupled with the respiratory chain via the flavin nucleotides, while the point of coupling of others, such as choline, is unknown.

The links of the respiratory chain are not yet fully established (for review, see Chance and Smith 1952) and the enzymes which it has been possible to isolate are often considered to be fragmentary and not identical with those bound to the structure. The intact system is represented by the classical heart muscle preparation of Keilin and Hartree. This is a particulate system which catalyses all oxidative stages between succinate or DPNH$_2$ and oxygen, but is practically devoid of the enzymes catalysing the reduction of pyridine nucleotides by different substrates. Since the succinic oxidase system and the DPNH$_2$ oxidase system are located in the mitochondria, the Keilin heart preparation can be regarded as possessing a level of organisation derived from the mitochondria (Green 1951 b). It was with this system that most of the earlier findings with regard to the cytochrome system were made. After very detailed and brilliant studies of the Keilin complex and of an analogous preparation from kidney, Slater (1948, 1949 a,

b, c, 1950 a, b) has in recent years accumulated evidence of a link between
the diaphorase and succinic oxidase systems. He has suggested the following scheme for the two oxidative pathways:

$$\text{DPNH}_2 \longrightarrow \text{flavoproteins} \searrow$$
$$\text{succinic dehydrogenase} \rightleftharpoons \text{cyt. } b \rightleftharpoons \text{``factor''} \to \text{cyt. } c \to \text{cyt. } (a\text{---}a_3) \to \text{O}_2$$

SLATER's factor is a porphyrin derivative, the nature of which is at present
unknown. It is characterized by a sensitivity to 2, 3-dimercaptopropanol (BAL)
and probably also to antimycin a (POTTER and REIF 1951). While the factor is
an obligatory link in the heart preparation, it is apparently not essential for the
activity of the purified enzyme DPNH$_2$-cytochrome c reductase (EDELHOCH,
HAYAISHI and TEPLY 1952) which catalyses the oxidation of DPNH$_2$ by cytochrome c. It is a matter of doubt, however, whether this enzyme has the same
activity in the isolated state as in the heart preparation (SLATER 1950 b). The
long-known enzyme diaphorase, which catalyses the oxidation of DPNH$_2$ by
methylene blue, but not by cytochrome c, is thought to be an artefact (POTTER
1950). Since the isolated DPNH$_2$-cytochrome c reductase is not inhibited by
antimycin a (EDELHOCH, HAYAISHI and TEPLY 1952), it would appear that this
also is not an intact enzyme but the two end components of a three-component
system, the central component, that carrying the factor, being lost in the preparation.

In contrast to the DPNH$_2$-cytochrome c system, the reaction succinate $\to$ factor
is considered to be reversible, which would explain the early observation that
fumarate oxidises DPNH$_2$ with the formation of succinate (SLATER 1950 b).
TPNH$_2$-cytochrome c reductase has been isolated in a purified form by HORECKER
(1950). It has not as yet been determined whether the hydrogen transport from
TPNH$_2$ to oxygen occurs in the same manner as applies to DPNH$_2$ (SLATER 1950 b).

The cytochrome c reductases are not exclusively located in the mitochondria. Rather do they appear to be concentrated in the microsomes
(HOGEBOOM 1949, HOGEBOOM and SCHNEIDER 1950 b, BRODY, WANG and BAIN
1952). Cytochrome oxidase, which is believed to be identical with cytochrome a_3 (cf. CHANCE and SMITH 1952), is exclusively located in the mitochondria (SCHNEIDER and HOGEBOOM 1951).

HUENNEKENS and GREEN (1950) have observed that if the mitochondria
suffer a loss of their coenzyme complement through ageing, their oxidative
activity can be restored by addition of coenzymes. In this case, however,
DPN can replace TPN, even in those oxidations which specifically require
TPN in purified systems. It has now been demonstrated (COLOWICK 1951 a)
that this apparent discrepancy is due to the presence in the mitochondria
of an enzyme catalysing the reaction:

$$\text{DPNH}_2 + \text{TPN} \to \text{TPNH}_2 + \text{DPN}$$

This reaction has proved to be not freely reversible, but to proceed,
strangely enough, only in the direction indicated above. The enzyme has
been isolated by COLOWICK, KAPLAN, NEUFELD and CIOTTI (1952) and experiments with it indicate (KAPLAN, COLOWICK and NEUFELD 1952) that the
reaction is far more complicated than was initially believed. It appears
to involve a transamination between the two nucleotides and their deamino
forms. The latter have been shown to replace the usual forms in various

dehydrogenase systems and may therefore be regarded as constituting a new type of hydrogen-transferring coenzymes.

B. Yield of Energy During Oxidations

1. Thermodynamic data

A general condition for the production of an energy-rich bond in an oxidation-reduction system is that the reaction:

$$Red_1 + Ox_2 \rightarrow Red_2 + Ox_1$$

represenst a fall in potential corresponding to the amount of energy included in the energy-rich bond. The composition of oxidation-reduction systems occurring in the living cell have therefore been studied by thermodynamic procedures (for numerical data, see Kaplan 1951 a) and it has been possible to identify the individual stages where this condition is fulfilled. These are (cf. Hunter 1951):

1. Oxidation of pyruvate and a-ketoglutarate by DPN.
2. Oxidation of $DPNH_2$ or $TPNH_2$ by FAD.
3. Oxidation of $FADH_2$ by cytochrome c.
4. Oxidation of red. cytochrome c by cytochrome oxidase.
5. Oxidation of cytochrome oxidase by O_2.

Accordingly, the oxidation by way of the Krebs cycle of one molecule of pyruvate to carbon dioxide and water, would theoretically involve the following oxidative steps in which high energy bonds are generated:

$$Pyruvate + oxaloacetate + H_2O + DPN \rightarrow citrate + CO_2 + DPNH_2 \qquad 1\ bond$$
$$DPNH_2 + \tfrac{1}{2}\,O_2 \rightarrow DPN + H_2O \ \ldots \ldots \ldots \ldots \ 4 \ ,,$$
$$Citrate + TPN \rightarrow a\text{-ketoglutarate} + CO_2 + TPNH_2 \ \ldots \ldots \ 0 \ ,,$$
$$TPNH_2 + \tfrac{1}{2}\,O_2 \rightarrow TPN + H_2O \ \ldots \ldots \ldots \ldots \ 4 \ ,,$$
$$a\text{-ketoglutarate} + DPN \rightarrow Succinate + CO_2 + DPNH_2 \ \ldots \ldots \ 1 \ ,,$$
$$DPNH_2 + \tfrac{1}{2}\,O_2 \rightarrow DPN + H_2O \ \ldots \ldots \ldots \ldots \ 4 \ ,,$$
$$Succinate + \tfrac{1}{2}\,O_2 + H_2O \rightarrow malate \ (via \ cyt. \ c) \ \ldots \ldots \ 2 \ ,,$$
$$Malate + DPN \rightarrow oxaloacetate + DPNH_2 \ \ldots \ldots \ldots \ 0 \ ,,$$
$$DPNH_2 + \tfrac{1}{2}\,O_2 \rightarrow DPN + H_2O \ \ldots \ldots \ldots \ldots \ 4 \ ,,$$

$$Pyruvate + 2\tfrac{1}{2}\,O_2 \rightarrow 3\,CO_2 + 2\,H_2O \ \ldots \ldots \ldots \ldots \ 20\ bonds$$

The experimental values hitherto obtained have, however, never exceeded 15, even if corrections for possible secondary losses of bonds formed are taken into consideration (cf. Hunter 1951).

Obviously the liberation in an oxidative step of an amount of energy corresponding to that of a high-energy bond, is not a sufficient condition for the actual formation of the bond.

A further requirement is that a mechanism shall be present which, when coupled with the oxidative process, will conserve the energy.

2. The P/O ratio

As these oxidative processes, in their capacity of generators of energy-rich bonds, have as yet only been amenable to study in intact systems, research has been confined to a comprehensive study of individual reactions in systems limited by the use of artificial hydrogen acceptors or anaerobic dismutations. A detailed review of this matter has recently been published by HUNTER (1951). The technique employed has been the determination of the number of energy-rich bonds synthesized in connection with a known number of electron transfers. The measure of comparison is the ratio of the number of micromoles of esterified phosphate to the number of electron pairs liberated, calculated as microatoms of oxygen (P/O ratio).

a) α-Keto acids

Pyruvate was the first substrate with which the P/O ratio was studied. In his study of intact systems, OCHOA (1943) already came to the conclusion that the P/O ratio for pyruvate was 3. In this system the pyruvate was completely oxidized to carbon dioxide and water by way of the Krebs cycle. Nevertheless, if the process is inhibited at the citrate stage, a value $P/O = 3$ is also obtained. Since the energy in the "active" acetate formed here is used in the formation of citrate by condensation with oxaloacetate (OCHOA 1952), we have for the transition $DPNH_2 \rightarrow O_2$ a P/O ratio of 3.

In bacterial systems the "active" acetate can be converted to acetyl phosphate where the energy is retained in the phosphate-bound form (LIPMANN 1939, 1940). In such systems, a P/O ratio of 4 is thus to be excepted (HUNTER 1951). In animal systems the "active" acetate cannot incorporate orthophosphate, but its energy can be transformed, alternatively to the condensation reaction, to phosphate-bound energy with the aid of inorganic pyrophosphate (LIPMANN, JONES, BLACK and FLYNN 1952, LIPMANN, JONES and BLACK 1952). This mechanism is of special interest in this connection, in that it gives a ratio of 3, if expressed as P/O; if, however, the ratio is expressed as the number of energy-rich bonds produced per oxygen atom, the value 4 is obtained. It should further be observed that if both the condensation reaction and the pyrophosphorolysis take place, as they presumably do in intact systems, a P/O ratio in the form of a whole number is extremely improbable.

The occurrence of a phosphorylation at the substrate level was first shown by HUNTER (1949) in his studies on the oxidation of α-ketoglutarate in the dismutation system

$$\alpha\text{-ketoglutarate} + \text{oxaloacetate} \rightarrow \text{succinate} + \text{malate} + CO_2$$

In this system it was possible to exclude the occurrence of a phosphorylation at the stage malate $\rightarrow$ oxaloacetate (LEHNINGER 1949). This has now been confirmed by a direct study of the isolated ketoglutarate oxidase (p. 51).

The process α-ketoglutarate $\rightarrow$ succinate differs from the process pyruvate $\rightarrow$ citrate, in that the phosphorolysis of succinyl-CoA is not an alternative pathway. The P/O ratio is therefore 4. It is not clear, however, why a P/O ratio approaching 4 is obtained when the intact system is provided with α-ketoglutarate as substrate (Lardy, Copenhaver and Wellman 1951).

It has been suggested (Green 1951 b) that since the other substrates produced in the degradation of ketoglutarate over the Krebs cycle have an average P/O ratio of two to three, the stage α-ketoglutarate $\rightarrow$ succinate should give a ratio of 5. This interpretation is, in our opinion, difficult to accept, before the general problem has been considered as to how the P/O ratio of the entire system can be dominated by an added substrate. It is known that addition of succinate as substrate gives rise to an average P/O ratio lower than that for any other substrate in the Krebs cycle. This may appear unreasonable in that all these substrates pass through the cycle an equal number of times, whichever of them is used as the initial substrate. We shall return to this matter when dealing with the oxidation of succinate.

b) Hydroxy acids

The two reactions isocitrate $\rightarrow$ oxalosuccinate and malate $\rightarrow$ oxaloacetate have been difficult to study in one-stage systems, since suitable inhibitors such as arsenite (Ochoa 1947) also affect the phosphorylation mechanism (Lehninger 1949, Hunter 1951). The only hydroxy acid with which unexceptionable one-stage investigations can be made is β-hydroxybutyrate, which only gives rise to acetoacetate on oxidation in liver systems. Since, on theoretical grounds, this reaction cannot give phosphorylation at the substrate level, it has proved useful in studies of the phosphorylation in the process $DPNH_2 \rightarrow O_2$ (Lehninger 1949, Lehninger and Smith 1949) (p. 44).

c) Glutamic acid

The oxidative deamination of glutamate, a reaction catalysed by DPN, has been studied in detail by Hunter and Hixon (1949) in the dismutation system:

(1) α-ketoglutarate $+ NH_3 + DPNH_2 \rightleftharpoons$ glutamate $+$ DPN
(2) α-ketoglutarate $+$ DPN $\rightarrow$ succinate $+ CO_2 + DPNH_2$

2 α-ketoglutarate $+ NH_3 \rightarrow$ glutamate $+$ succinate $+ CO_2$

This reaction has been found to give a phosphorylation which, as has now been demonstrated, is to be attributed to reaction (2). Hence no energy-rich bonds are generated in the oxidative deamination of glutamate, which is in accordance with thermodynamic calculations (Hunter and Hixon 1949 a).

d) Succinic acid

The succinic oxidase preparation of Keilin and Hartree has not the ability of esterifying phosphate in connection with the oxidation. In more intact systems, however, it has been possible at an early stage to obtain direct evidence for a phosphorylation associated with the hydrogen transport between succinate and oxygen (Colowick, Welch, Cori 1940). It has, however, been difficult in these experiments to prevent the further oxidation of the fumarate. Arsenite is an unsuitable inhibitor for this purpose, since it may also inhibit phosphorylation (HUNTER 1951). OCHOA (1943), CROSS, TAGGART, COVO and GREEN (1949), EILER and McEWEN (1949) and HUNTER and HIXON (1949) have obtained P/O ratios of between 1 and 2 in uninhibited systems, where they were nevertheless able to show that the oxidation of fumarate took place at only a fraction of the rate at which the succinate was oxidized.

This again shows the danger of deriving values for the individual stages from the average P/O ratios. It is obviously necessary to reckon with the fact that the P/O ratio for the added substrate can dominate the rest of the system. This fact is probably due to a limiting factor for the mitochondrial respiration in the cytochrome region (HOGEBOOM 1949, HOGEBOOM and SCHNEIDER 1950 b. Cf. also POTTER, RECKNAGEL and HURLBERT 1951 and discussion on p. 93), so that a substrate present in relatively high concentration will enjoy a "priority" over the rest of the system. Clear-cut evidence in favour of such reasoning is to be found in the paper of LEUTHARDT and MAURONS (1950) who find that succinate is oxidized in mitochondria as rapidly in the absence as in the presence of phosphate, despite the fact that in the former case only the one-stage reaction succinate→fumarate takes place, whereas in the latter the whole cycle is involved. In the light of these considerations, the above-mentioned values of 1 to 2 for the P/O ratio may be taken as characteristic of the stage succinate→fumarate.

Recently more direct evidence has come to light for the occurrence of phosphorylation between succinate and fumarate. HERSEY and AJL (1951 a, b) have obtained an enzyme preparation from *Escherichia coli* which catalyses phosphorylation coupled with hydrogen transfer from succinate to oxygen. GREEN, BEINERT, FULD, GOLDMAN, PAUL and SARKAR (1953) have independently isolated from pig's heart an insoluble enzyme system which catalyses the same reaction (Table 5). In both cases the enzymes could not oxidize fumarate. It is uncertain whether phosphorylation occurs at the actual

Table 5. *Oxidative phosphorylation in the oxidation of succinate.*
(From GREEN. BEINERT. FULD. GOLDMAN. PAUL and SARKAR. 1953.)

	Phosphate esterified (μ moles)	Oxygen absorbed (μ atoms)	P/O
Complete system	25.2	78.5	0.38
no ATP	0.1	91.3	0.00
no hexokinase	1.6	79.9	0.02
no succinate	0.0	0.0	0.00
no magnesium	9.8	81.5	0.13
with 0.0002 M 2, 4-dinitrophenol	0.0	69.7	0.00

The complete system contained 0.3 ml of "5.4" enzyme. 8 μ moles ATP, 80 μ moles phosphate (pH 7.4). yeast hexokinase (2.5 mg). 100 μ moles fructose, 150 μ moles succinate, and 2 μ moles MgCl$_2$. Final vol. 3.0 ml. Time 70 min.

substrate level. This is, however, scarcely conceivable, since the reactions succinate→cyt b→Slater's factor are freely reversible in the presence of $DPNH_2$ and diaphorase (Slater 1950 b). It is hence probable that one or two phosphorylations occur in the stages between Slater's factor and oxygen.

e) Phosphorylations coupled to the electron transfer system

Oxidative phosphorylation between $DPNH_2$ and oxygen has been intensively studied by Lehninger, who has also written a review of the matter (1951). This author investigated, in three different systems, the stages at which phosphorylation occurs in mitochondria when $DPNH_2$ is oxidized by oxygen.

The most valuable results from the quantitative viewpoint were obtained in experiments where β-hydroxybutyrate served as substrate for liver mitochondria. Since this system cannot metabolize the acetoacetate formed, this substance accumulates in a quantity equivalent to the oxygen consumed. The liberated electrons pass over DPN, FAD and the cytochrome system to oxygen. Since, as has been noted above, no phosphorylation at the substrate level takes place in the oxidation of hydroxy acids, all the phosphorylation found in Lehninger's system must be associated with the electron transport between $DPNH_2$ and oxygen. The P/O ratios obtained varied between 1.30 and 2.44. When these values were corrected for loss of phosphate, due to phosphatases, a probable value of 3 was obtained.

In more recent experiments Lehninger has confirmed his results regarding the occurrence of phosphorylation between $DPNH_2$ and oxygen. For one series he employed substantial quantities of highly purified $DPNH_2$ and in a second series he added to the mitochondrial suspension $DPNH_2$-generating systems such as alcohol dehydrogenase and phosphoglyceraldehyde dehydrogenase. These systems gave rise to phosphorylation only when the mitochondria, by hypotonic treatment, were made accessible to the extraneous $DPNH_2$. It also proved necessary to add cytochrome c to the system. In view of these special circumstances, it was not possible to derive quantitative data for the P/O ratio from the experiments.

Lardy, Copenhaver and Wellman (1951) employed hexokinase and glucose in conjunction with the same system in order to tap off the high-energy phosphate as it was formed. This procedure is ideal, since it not only excludes loss of esterified phosphate, but also ensures constant respiration and phosphorylation during the experiment, and thus more reproducible results. In this manner Lardy and co-workers obtained a value between 2.7 and 3.

Attempts to locate the single stages between $DPNH_2$ and oxygen associated with phosphorylation have met with small success and the results are, indeed, somewhat conflicting. The investigations have largely been concentrated on the problem whether phosphorylation occurs between cytochrome c and oxygen.

Two techniques have been utilized. One of these involves an uncoupling of the electron transport between cytochrome c and oxygen.

Slater (1950 c), in his studies on the oxidation of α-ketoglutarate, employed considerable quantities of cytochrome c as electron acceptor, inhibiting the oxidation of the cytochrome with cyanide, or with absence of oxygen. In these experiments a P/O ratio of between 0.98 and 2.20 was found. In the oxidation of α-ketoglutarate with oxygen in the same system the value varied between 1.67 and 2.79. The similarity between the two groups of values can scarcely be taken as conclusive evidence that no phosphorylation takes place between cytochrome c and oxygen. As Lehninger (1951) points out, moreover, the values for oxidation and phosphorylation in Slater's anaerobic experiments were too small to permit any final conclusions. It may be added here that the account of the experiments gives no direct evidence to exclude the possibility that a myokinase effect contributed to the formation of glucose-6-phosphate from the added ADP. Finally, it may be remarked that, as Hunter (1951) has suggested, cytochrome c may behave differently when it is introduced into the system and when it functions within the mitochondria.

Other authors (Loomis and Lipmann 1948, Cross, Taggart, Covo and Green 1949, McEwen and Eiler 1950) have attempted to uncouple the cytochrome system from the respiratory chain with aid of ferricyanide. It is not known with certainty, however, at which stage ferricyanide inhibits the system; nor has the toxicity of this substance for the phosphorylating mechanism been established. It is therefore difficult to draw conclusions from these experiments (Hunter 1951).

The second group of experiments, designed to elucidate the status of phosphorylation in the cytochrome system, involved the use of substrates which are directly oxidized by the cytochrome system. These experiments were made by Friedkin and Lehninger (1949 b). No phosphorylation was found when glutathione, cysteine, hydroquinone and p-phenylenediamine were employed as substrates. Some phosphorylation was, however, obtained with ascorbic acid. This phosphorylation was low in comparison with that obtained, for example, with malate, but a control experiment showed that the phosphorylation with malate was also reduced in the simultaneous presence of ascorbic acid. It was subsequently shown by Bernheim, Wilbur and Kinaston (1952) that ascorbic acid had an inhibitory effect on certain oxidative enzyme systems in the mitochondria and that this effect was associated with the appearance of unsaturated fatty acids from a lipid fraction. In view of these facts it seems that the phosphorylation obtained by Friedkin and Lehninger with ascorbic acid is of some significance, despite its low value. On the other hand, as the authors themselves have pointed out, it is difficult to decide whether the phosphorylation actually takes place between cytochrome c and oxygen, since nothing is known concerning the fate of the dehydroascorbic acid formed in the reaction.

By way of conclusion, it may be said that there is no longer any doubt that phosphorylations occur in the respiratory chain between $DPNH_2$ and oxygen; there is much evidence, indeed, that the greater part of the phosphorylation occurs in *this* region. Thermodynamic calculation shows that four phosphorylations are theoretically possible, whereas only three have been detected experimentally, even when reasonable corrections are made. It has, however, proved a matter of the greatest difficulty to find suitable experimental conditions for studies of phosphorylation in this part of the respiratory system. For the same reason, it has been difficult to associate the phosphorylations with the individual electron steps. It cannot, at present, be excluded that phosphorylations may take place within the cytochrome system.

C. Mechanism of Formation of Energy-Rich Bonds

1. Hypotheses based on phosphorylation prior to oxidation

The earliest hypothesis for the mechanism of energy-binding in phosphate bonds was presented by Warburg and his co-workers at the end of the 1930's. Warburg's school had been concerned with the reaction mechanism for the oxidation of 3-phosphoglyceraldehyde. The overall reaction catalysed by the crystalline enzyme (Warburg and Christian 1939, Cori, Slein and Cori 1948, Cori, Velick and Cori 1950) may be written:

$$3\text{-phosphoglyceraldehyde} + \text{DPN} + \text{orthophosphate} + \text{ADP}$$
$$\rightarrow 3\text{-phosphoglyceric acid} + \text{DPNH}_2 + \text{ATP}$$

Since the oxidation here requires the presence of orthophosphate, it was postulated that phosphate was first bound in ester linkage to a hydrated phosphoglyceraldehyde molecule and that this ester linkage, after oxidation, was converted to an energy-rich acyl phosphate bond. Of the two hypothetical intermediates, 1,3-diphosphoglyceraldehyde and 1,3-diphosphoglyceric acid, the latter was detected and isolated by Negelein and Brömel in 1939. No evidence for the occurrence of 1,3-diphosphoglyceraldehyde has, however, yet been obtained.

Nevertheless, the overwhelming majority of the workers engaged in the study of phosphorylation mechanisms in other systems, have accepted the general validity of the principle proposed by Warburg, namely that the substrate, before oxidation, could spontaneously add phosphate, and that this would be converted by oxidation into an energy-rich acyl phosphate.

This concept received strong support from the early studies by Lipmann (1946a) of the acetylation mechanism. This worker was at the time interested in the oxidation of pyruvic acid and suggested that the pyruvate, before undergoing oxidation, coupled with phosphate to give a phosphorylated, semiacetal-like compound. This, after oxidative removal of carbon dioxide or formate, would yield the energy-rich acetyl phosphate. The demonstration of the formation of acetyl phosphate and the isolation of the compound (Lipmann 1946a) tended to confirm Lipmann's reasoning, just in the same way as Warburg had confirmed his hypothetical scheme for the oxidation of glyceraldehyde by the isolation of 1,3-diphosphoglyceric acid. In this case also, evidence for the occurrence of the hypothetical substrate of the oxidation was lacking.

Lipmann (1946 b) nevertheless considered that his evidence sufficed for the formulation of a more general theory of the formation of energy-rich acyl phosphate. This provided, above all, for the formation of phosphate-bound energy at the coenzyme level. According to this hypothesis, later presented in a more elaborate form by Kaplan (1951 b), Lipmann assumed that a hypothetical keto group acted as an intermediate acceptor of electrons between the pyridine and the flavin nucleotides. During the electron transport this hypothetical intermediate was converted into a seconary alcohol. This was coupled with phosphate and after re-oxidation was converted to an energy-rich enol phosphate. Kaplan has extended this hypothesis, suggesting that the hypothetical keto group is that of oxaloacetate, which, by way of phosphomalate, could give rise to phospho*enolo*xaloacetate.

It became difficult, however, to maintain this theory when the Lehninger group showed that synthetic phosphomalate could not act as an intermediate in the oxidation of fumarate (Friedkin and Lehninger 1947) and that the oxidation of $DPNH_2$ was associated with phosphorylation in a system which was demonstrably free from dicarboxylic acids (Lehninger 1949).

More recently, Kaplan has also discarded his assumption of a dicarboxylic intermediate. In a series of experiments (Kaplan 1951 b), he has made use of the well-known phenomenon that cyanide can be bound reversibly to the α-C-atom of the pyridine nucleus of DPN. This produces a change in the absorption spectrum of the coenzyme, analogous to that occurring on reduction. From this, and a number of other observations, Kaplan has arrived at the conclusion that the C=N linkage in the pyridine nucleus strongly resembles an aldehyde configuration. Addition of phosphate to the hydrated form of the nucleus would thus result in a low-energy ester phosphate linkage, which after oxidation would be converted to an energy-rich carbonyl phosphate bond. When this carbonyl phosphate group was transferred to ADP, an oxidized form of DPN, a pyridone, would result. This would dismute with a molecule of reduced DPN to give two molecules of DPN. Kaplan's hypothetical scheme is shown in Fig. 14. The reactions in the scheme may be summarized thus:

$$DPN + phosphate \rightleftharpoons DPN\text{-}phosphate$$
$$DPN\text{-}phosphate + FAD \rightarrow pyridone \sim phosphate + FADH_2$$
$$pyridone \sim phosphate + ADP \rightleftharpoons pyridone + ATP$$
$$pyridone + DPNH_2 \rightleftharpoons 2\,DPN$$

Net: $DPNH_2 + FAD + phosphate + ADP \rightarrow DPN + FADH_2 + ATP$

It will be noted that all these hypotheses are based on a pre-phosphorylation of a substrate and the occurrence of a carbonyl phosphate for each coupling reaction between oxidation and phosphorylation. Of such carbonyl phosphates it has been possible, in bacterial systems, to demonstrate, in addition to acetyl phosphate, a number of phosphorylated fatty acids as active intermediates (Lipmann 1946 a, Barker 1952). Succinyl phosphate is one of the carbonyl phosphates whose existence in these systems is still postulated (AJL and WERKMAN 1948, OCHOA 1951).

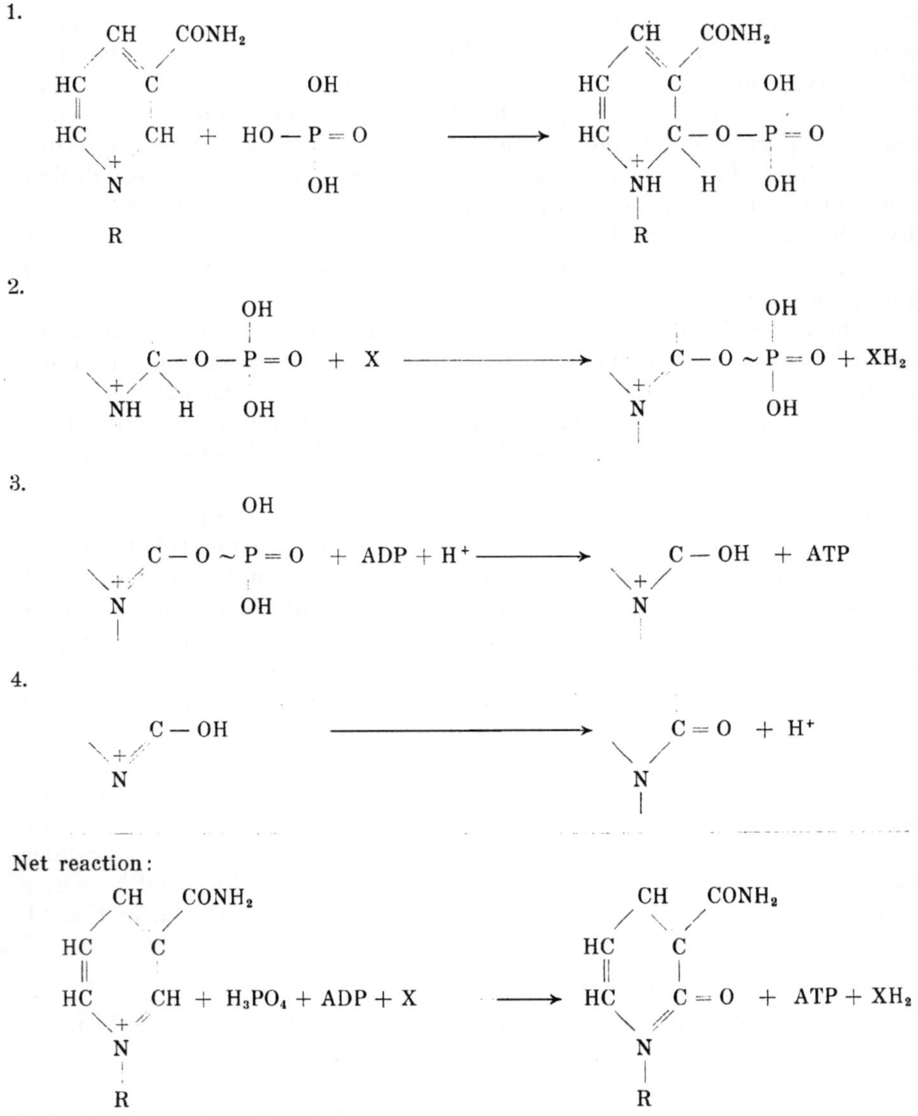

1.

$$\text{pyridinium nucleotide} + HO-P=O(OH)_2 \longrightarrow \text{phosphorylated dihydropyridine adduct}$$

2.

$$C-O-P=O(OH) + X \longrightarrow C-O\sim P=O(OH) + XH_2$$

3.

$$C-O\sim P=O(OH) + ADP + H^+ \longrightarrow C-OH + ATP$$

4.

$$C-OH \longrightarrow C=O + H^+$$

Net reaction:

$$\text{pyridinium nucleotide} + H_3PO_4 + ADP + X \longrightarrow \text{oxidized nucleotide} + ATP + XH_2$$

Fig. 14. Hypothetical mechanism of phosphorylation coupled to the transfer of electrons by pyridine nucleotides. (According to KAPLAN 1951 b.)

Such phosphorylated products, including acetyl phosphate (LIPMANN 1946 a), have not, on the other hand, been demonstrated in animal tissues. On the basis of the experimental data now available their occurrence can, indeed, be decisively excluded. It is of historical importance that it was precisely this failure of animal tissue to fit into the hypothetical scheme, which led to what would seem to be a more definitive explanation of the formation of phosphate-bond energy.

2. Generation of energy-rich thiol bonds

Our present conception of the mechanism of generation of energy-rich bonds originates from studies on the problem of acetylation in animal tissues. It was clear at an early stage that acetyl phosphate could not serve as energy donor in acetylating systems of animal origin, such as LIPMANN's (1946 a) liver system for the acetylation of sulphanilamide or the system, studied by NACHMANSOHN and MACHADO (1943), which acetylated choline. In these experiments the important discovery was made that in the transfer of an acetyl group to choline in the presence of ATP as energy source, a thermostable co-factor, subsequently named coenzyme A, was involved.

In later work, the constitution of CoA was determined, the result of a brilliant collaboration between a number of workers (KAPLAN and LIPMANN 1948, NOVELLI, KAPLAN and LIPMANN 1949, SNELL, BROWN, PETERS, CRAIG, WITTLE, MOORE, McGLOHON and BIRD 1950, NOVELLI 1951, BADDILEY and THAIN 1951, LYNEN and REICHERT 1951, LYNEN, REICHERT and RUEFF 1951, GREGORY and LIPMANN 1952, BADDILEY, THAIN, NOVELLI and LIPMANN 1953), the coenzyme was isolated (LIPMANN, KAPLAN, NOVELLI, TUTTLE and GUIRARD 1950, BEINERT, VON KORFF, GREEN, BURKER, HANDSCHUMACHER, HIGGINS and CHONG 1952), and reactions involving CoA were studied (for review, see LIPMANN, JONES and BLACK 1952). These studies, and especially the important contributions of the LYNEN and OCHOA groups (STERN, OCHOA and LYNEN 1952) regarding the energetics of the derivatives of CoA, have led to a very extensive revision of earlier hypotheses of the origin of energy-rich bonds. The development of a more or less definitive picture first became possible, however, with the solubilization and isolation of the α-keto acid oxidases.

a) Solubilization of mitochondrial oxidases

Attempts to bring the oxidative enzymes of the mitochondria into solution have been made with the aid of various techniques. The OCHOA school employed homogenization with the Waring blendor to destroy the particulate structure of pig's heart. From the supernatant obtained after high-speed centrifugation of such homogenates it was possible to make enzyme preparations catalysing the initial stages of the oxidations of pyruvate (OCHOA 1952) and of α-ketoglutarate (KAUFMAN 1952). The findings made regarding the mechanism of these oxidases are dealt with later under the relevant headings.

Another procedure for the release of enzymes bound to the particulate structure has been employed by DRYESDALE (1952) who, by extraction and subsequent ammonium sulphate fractionation of acetone-dried liver mitochondria, obtained an enzyme preparation which catalyzed the oxidation of fatty acids to acetoacetate in the presence of dichlorophenol-indophenol as hydrogen acceptor.

A more general technique for the liberation of mitochondrial oxidases has been described by Green, Beinert, Fuld, Goldmann, Paul and Sarkár (1953). By a combination of cyclophorase isolation and ammonium sul-

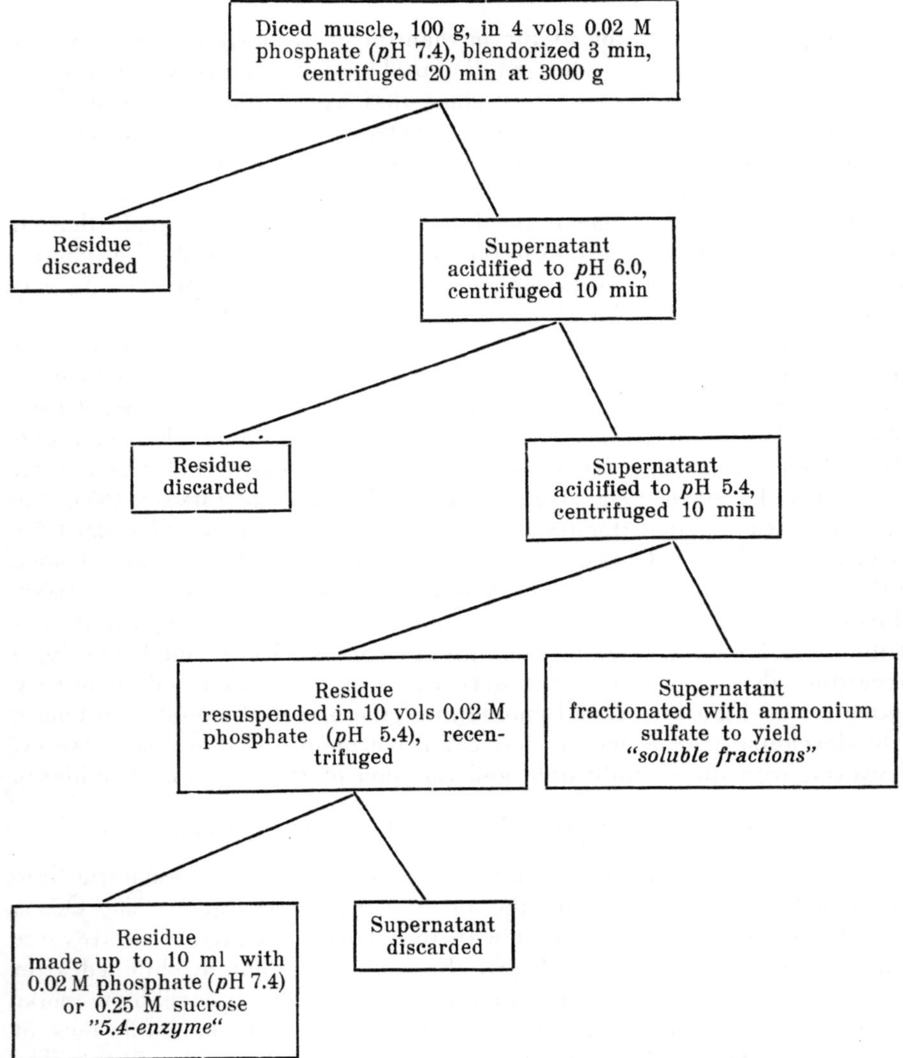

Fig. 15. Preparation of particulate and soluble fractions of pig heart muscle.
(According to Green, Beinert, Fuld, Goldman, Paul and Sarkar 1953.)

phate fractionation applied to pig heart homogenates (see Fig. 15), these authors have obtained two enzyme fractions, an insoluble one, called "5.4 enzyme", and a soluble "supernatant".

These two fractions together, supplemented with a coenzyme concentrate prepared from yeast or liver, containing the usual complement of coenzymes, catalysed the complete degradation to carbon dioxide and water of all the intermediates of the Krebs cycle with the exception of oxaloacetate, for the oxidation of which additions of pyruvate and ATP were necessary. The oxidation of acetoacetate by this system required "sparking" in the form of a Krebs cycle intermediate, while that of acetate also required ATP. The oxidation of fatty acids showed the same requirements as did that of acetate, although the higher fatty acids were not oxidized by this system.

The presence of the supernatant fraction was necessary for all substrates with the exception of succinate, which could be oxidized to fumarate by the "5.4 enzyme" alone. If this latter enzyme were supplemented with ATP and hexokinase, it was possible to obtain phosphorylation coupled with the oxidation of succinate to fumarate. A P/O ratio of about 0.4 was found. This constitutes strong evidence for the occurrence of phosphorylation in the cytochrome system (cf. p. 45). By combination of the "5.4 enzyme" with a supernatant fraction which was in some degree freed from glycolytic enzymes, it was possible to demonstrate phosphorylation in connection with the two-stage oxidation of α-ketoglutarate to fumarate; the P/O ratio was 0.5.

From these and other observations, the authors concluded that the "5.4 enzyme" contained the succinic oxidase system including the cytochrome components, together with the phosphorylating apparatus associated with this system. The supernatant fraction, on the other hand, contained the oxidases for all the remaining Krebs cycle intermediates and for the fatty acids, the DPN-cytochrome c reductase, the activating principle for acetate and fatty acids, and the phosphorylating principles associated with the respective oxidases. With the aid of artificial hydrogen acceptors, the oxidation of different substrates could be catalysed by the supernatant fraction alone. It was from this fraction that the pyruvic oxidase and the α-ketoglutaric oxidase were subsequently isolated.

b) The α-keto acid oxidases

The oxidation of pyruvate in purified enzyme systems has been studied most extensively by Ochoa, Korkes, del Campillo and Gunsalus (Ochoa 1952) and by Schweet, Fuld, Cheslock and Paul (1951), Jagannathan and Schweet (1952), Schweet, Katchman, Bock and Jagannathan (1952), Schweet and Cheslock (1952). These investigations were concerned with the mechanism of the formation of acetyl-CoA from pyruvate.

By the fractionation of bacterial extracts, the Ochoa group isolated an enzyme complex which, in the presence of cocarboxylase, DPN and CoA, converted pyruvate to carbon dioxide and "active" acetate. The overall reaction may be written:

$$\text{pyruvate} + \text{DPN} + \text{CoA} \rightarrow \text{acetyl-CoA} + \text{DPNH}_2 + \text{CO}_2$$

Since enzymes were already known which could decarboxylate pyruvate to acetaldehyde (carboxylase) and oxidize acetaldehyde to acetate (acetaldehyde dehydrogenase, *C. kluyveri*), it was reasonable to suppose that the reaction took place in several stages.

Attempts to demonstrate acetaldehyde in the reaction mixture were fruitless. The crude enzyme extract could, however, be separated into two fractions, termed A and B, which only together could catalyse the above overall reaction. The authors hence suspected that fraction A contained a decarboxylase and fraction B a dehydrogenase. It was suggested that the mechanism of the reaction was as follows:

(1) Pyruvate + X $\rightleftharpoons$ acetaldehyde-X + CO_2.

(2) Acetaldehyde-X + CoA $\rightleftharpoons$ acetaldehyde-CoA + X.

(3) Acetaldehyde-CoA + DPN $\rightleftharpoons$ acetyl-CoA + $DPNH_2$.

Reaction (1) was thought to be catalysed by enzyme A and reaction (3) by enzyme B. It was actually possible to demonstrate reaction (1) in the presence of enzyme A with only diphosphothiamine present as coenzyme, for which reason it has been suggested that X may be identical with diphosphothiamine. The reaction proceeds in the absence of CoA, and only when it is coupled to enzyme B does the presence of CoA become essential. The authors have more recently shown these two enzyme fractions to be present also in animal tissues.

While the above investigations were proceeding, the SCHWEET group was subjecting the pyruvic oxidase system to detailed study. These experiments were made with animal tissue, principally pigeon breast muscle.

In agreement with the OCHOA group, these authors also found that two protein components were necessary to catalyse the overall reaction mentioned above. One of these was isolated in the pure state. This has been called the pyruvic oxidase and it is identical in function with OCHOA's enzyme A. The enzyme is electrophoretically homogeneous and has a molecular weight of 4 million. The only coenzyme that it requires is cocarboxylase. The reaction product is protein-bound acetaldehyde. This two-carbon body has the property of condensing with added aldehydes with the formation of acyloins without itself passing through the free aldehyde stage. It is autoxidizable, being split by oxygen to protein and acetic acid. More effective oxidants, however, are ferricyanide, methylene blue and pyocyanine, the latter being inhibited by catalase.

With the aid of a protein fraction isolated by GREEN, BEINERT, FULD, GOLDMAN, PAUL and SARKAR (1953) from pig heart mitochondria by ammonium sulphate fractionation, this two-carbon compound is oxidized catalytically by DPN, with a protein-acetyl compound as reaction product. In this reaction a further thermostable compound is necessary. The second enzyme component is probably identical with OCHOA's enzyme B. It apparently catalyses the transfer of the activated acetyl group to CoA.

The authors postulate that the primary acetyl-protein compound is energy-rich and have shown that, in the presence of CoA, the acetyl group can be transferred to sulphanilamide without further provision of energy. The reaction sequence suggested by SCHWEET and co-workers is this:

$$PYRUVATE$$
$$-\ CO_2 \downarrow \text{cocarboxylase}$$
$$[CH_3 . CO^- -{}^+E] \xrightarrow{\ RCHO\ } RCH(OH) . CO . CH_3$$
$$-\ 2\,e \downarrow \text{coenzyme}$$
$$[CH_3 . CO^+ \sim {}^- E] \xrightarrow{\ CoA\ } ACETYL \sim CoA$$

The mechanism of the oxidation of α-ketoglutarate resembles in a high degree that of pyruvate. The system in the isolated soluble form has been studied principally by KAUFMAN (1951), SANADI and LITTLEFIELD (1951) and SANADI, LITTLEFIELD and BLOCK (1952).

KAUFMAN isolated from pig heart an enzyme system which catalysed the oxidative decarboxylation of α-ketoglutarate. The essential co-factors were cocarboxylase, CoA and DPN. The reaction products were succinate, CO_2 and $DPNH_2$. As CoA was essential, it was supposed that the reaction proceeded in two stages:

$$\alpha\text{-ketoglutarate} + CoA + DPN \rightarrow \text{succinyl-CoA} + CO_2 + DPNH_2$$
$$\text{succinyl-CoA} + H_2O \rightarrow \text{succinate} + CoA$$

The first stage represents the oxidation proper, while the second is a deacylase reaction.

Almost simultaneously with the investigations described above, SANADI and LITTLEFIELD isolated from the same starting material an electrophoretically homogeneous enzyme material with a molecular weight of about two million. This enzyme catalysed the oxidation of α-ketoglutarate to succinate and CO_2 in the presence of 2, 6-dichlorophenol-indophenol as hydrogen acceptor. Since the addition of co-factors proved to be unnecessary, it was attempted to detect CoA, DPN, cocarboxylase and α-lipoic acid in the preparation. Only the two latter substances were found: 1.2 moles of cocarboxylase and 6 moles of α-lipoic acid per mole of enzyme.

It was found that α-ketoglutarate disappeared under anaerobic conditions, although a corresponding formation of succinic semialdehyde or (in the presence of added aldehydes) of acyloin could not be demonstrated. It was nevertheless assumed that a reaction product at the oxidation level of an aldehyde was present.

In the presence of dichlorophenol indophenol, succinate accumulated. If, however, DPN and CoA were added to the system, the α-ketoglutarate was oxidized with the formation of equivalent amounts of $DPNH_2$ and succinyl-CoA (SANADI and LITTLEFIELD 1952). It has been possible to accumulate and isolate the succinyl-CoA formed in this reaction. Its constitution has been shown to be:

$$R_{CoA} - S \sim CO . CH_2 . CH_2 . COOH$$

which is analogous with the constitution of acetyl-CoA. The bond between the succinyl radical and CoA is energy-rich, since the succinyl group can be transferred enzymically to sulphanilamide.

In the presence of DPN in stoichiometric amounts with relation to the α-ketoglutarate, the reaction may be employed for the quantitative estimation of CoA by spectrophotometric estimation of the $DPNH_2$ formed (GERGELY, HELE, and RAMAKRISHNAN 1952). If CoA is only present in catalytic amounts, the reaction cannot proceed unless the succinyl-CoA is continuously decomposed either by phosphorolysis (KAUFMAN 1951, see also p. 60) or by hydrolysis catalysed by the enzyme deacylase, recently isolated by GERGELY and associates (1952). This enzyme brings about a hydrolytic breakdown of succinyl-CoA to succinate and CoA.

The thermostable factor mentioned in connection with the pyruvic oxidase system, is identical with the α-lipoic acid acting in the oxidation of α-ketoglutarate. This co-factor, which was found some years ago to be a growth factor for certain microorganisms, has been extensively studied by the GUNSALUS group (O'KANE and GUNSALUS 1947, 1948, REED, DE BUSK, GUNSALUS and HORNBERGER 1951, GUNSALUS, DOLIN and STRUGLIA 1951, GUNSALUS, STRUGLIA and O'KANE 1951. REED and DE BUSK 1952 a, b). It is a dithioctic acid with the $-SH$-groups in the 5, 8- or 6, 8-positions. It has been found to occur conjugated with one molecule of thiamine pyrophosphate. In the light of the information now available, the compound is lipothiamide pyrophosphate (LTPP). Its probable formula is given in Fig. 16.

The rôle of the coenzyme in the oxidation of α-ketoglutaric acid is, the authors suggest, that of primary acceptor for the active acyl group (see p. 61).

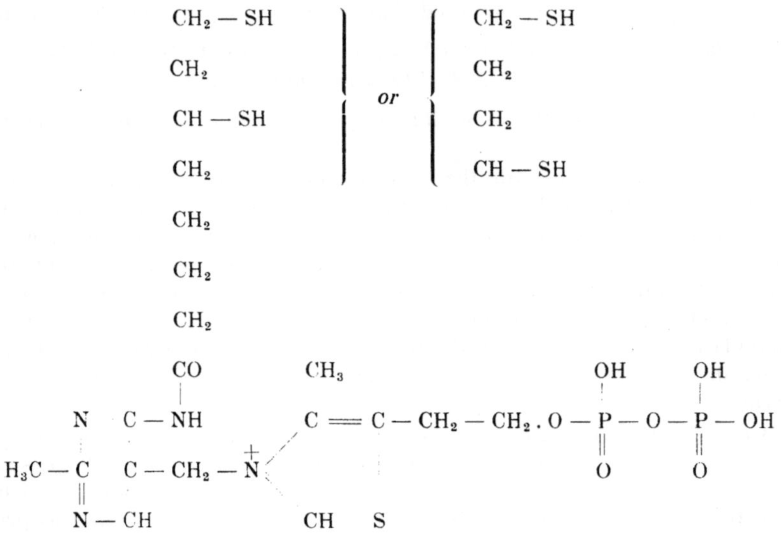

$$CH_2 - SH \qquad\qquad CH_2 - SH$$
$$CH_2 \qquad\qquad\qquad CH_2$$
$$CH - SH \qquad or \qquad CH_2$$
$$CH_2 \qquad\qquad\qquad CH - SH$$
$$CH_2$$
$$CH_2$$
$$CH_2$$

Fig. 16. Structure of lipothiamide pyrophosphate.
(According to available data.)

A general scheme for the oxidation of α-keto acids, consistent with the experimental findings outlined above, is shown in Fig. 17.

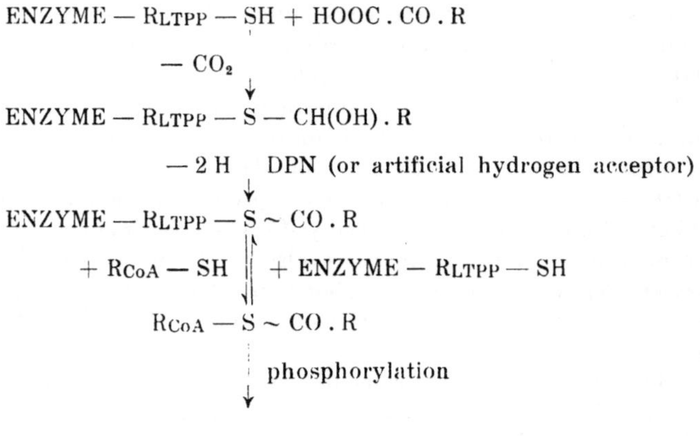

$$\text{ENZYME} - R_{LTPP} - SH + HOOC . CO . R$$
$$\downarrow - CO_2$$
$$\text{ENZYME} - R_{LTPP} - S - CH(OH) . R$$
$$\downarrow - 2 H \quad DPN \text{ (or artificial hydrogen acceptor)}$$
$$\text{ENZYME} - R_{LTPP} - S \sim CO . R$$
$$+ R_{CoA} - SH \quad + \text{ENZYME} - R_{LTPP} - SH$$
$$R_{CoA} - S \sim CO . R$$
$$\downarrow \text{phosphorylation}$$

Fig. 17. Oxidative decarboxylation of α-keto acids.

The studies in the oxidation of the α-keto acids and the findings concerning the energetics of the S-acyl and S-phosphoryl bonds, have led to the belief that the energy-rich bonds generated in the oxidation of these substrates, do not involve phosphate groups. According to this new concept, the formation of the energy-rich S-acyl linkage takes place in exactly

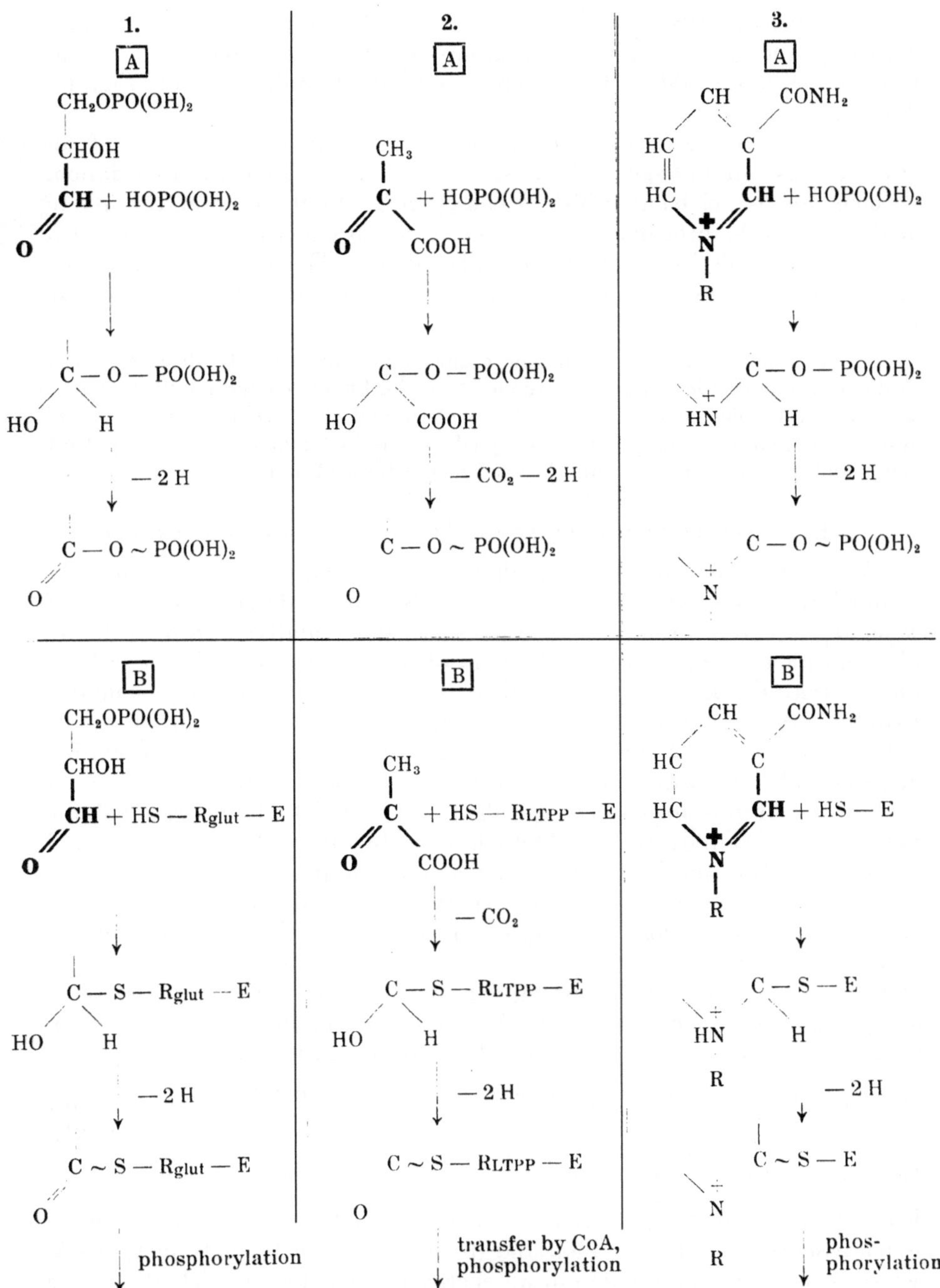

Fig. 18. Survey of hypotheses on the formation of phosphate-bound energy.

Series A: Hypotheses based on phosphorylation *prior to oxidation.* 1) Oxidation of 3-phosphoglyceraldehyde (according to WARBURG and CHRISTIAN 1939). 2) Oxidation of pyruvate (according to LIPMANN 1946 b). 3) Oxidation of DPN (according to KAPLAN 1951 b). *Series B:* Phosphate bonds formed *secondarily,* by transfer of energy from thiol bonds generated during oxidation. 1) Oxidation of 3-phosphoglyceraldehyde (according to RACKER and KRIMSKY 1952). 2) Oxidation of pyruvate (according to SCHWEET and co-workers 1952). 3) Oxidation of DPN (according to present work).

the manner originally postulated by Warburg, but with the difference that SH-compounds replace phosphate groups in the initial spontaneous combination with the aldehyde or keto group of the substrate (cf. Fig. 18, 2 A–B).

The initial stage of the process thus consists in the formation of an enzyme-substrate linkage of thiol-ester type. This linkage, after oxidation, becomes an S-acyl bond, which is energy-rich. In the case of the α-keto acid oxidases the prosthetic group may, as has previously been mentioned, be identical with lipothiamide pyrophosphate. The acyl group is then taken over by CoA, and it is only after this stage that phosphate enters the scene.

In the transfer reaction which now occurs, and which will be discussed later in detail, phylogenetic differences are clearly to be found: bacteria split acyl-CoA with the aid of phosphate to give acyl phosphate and CoA, whereas in the animal system the inorganic phosphate is coupled to the CoA-moiety of the acyl-CoA compound, the acyl group being liberated in the form of acid.

c) The 3-phosphoglyceraldehyde dehydrogenase

The recognition of the fact that phosphorylation is not primarily coupled with the oxidation of α-keto acids, has rendered it a matter of some urgency to reconsider the mechanism of the oxidative phosphorylation of 3-phosphoglyceraldehyde in the light of these findings. It has long been known that the molecule of the dehydrogenase concerned here contains two SH groups per molecule, and furthermore that, in addition to its natural substrate, the same enzyme is also able to catalyse the oxidation of glyceraldehyde, although at a considerably lower rate.

In view of this incomplete specificity, Harting (1951) investigated the possibility of employing acetaldehyde as substrate. Oxidation occurred in this case also, and despite the low reaction rate it was possible to isolate acetyl phosphate in significant amounts. This observation increased Harting's suspicion that phosphate was not primarily concerned in the reaction, in analogy with the oxidation of pyruvate. Consequently, Harting and Velick (1952) attempted to couple the system with transacetylase, condensing enzyme, CoA and oxaloacetate, and succeeded in obtaining citrate. This system may be represented as follows:

$$\text{I.}\quad \text{acetaldehyde} + \text{DPN} + \text{CoA} \xrightarrow{\text{dehydrogenase} + \text{transacetylase}} \text{acetyl-CoA} + \text{DPNH}_2$$

$$\text{II.}\quad \text{acetyl-CoA} + \text{oxaloacetate} \xrightarrow{\text{condensing enzyme}} \text{citrate} + \text{CoA}$$

From the result of this experiment it must be concluded: (a) that phosphate is not essential for the reaction, and (b) that reaction I yields acetyl-CoA, which involves the production of an energy-rich bond in the dehydrogenase reaction.

At about the same time, the Munich group (Holzer 1952, Holzer and Holzer 1952) independently suggested the following mechanism for the glyceraldehyde dehydrogenase reaction:

I. phosphoglyceraldehyde + enzyme-SH ⇌ enzyme-S-phosphoglyceraldehyde

II. enzyme-S-phosphoglyceraldehyde + DPN ⇌ enzyme-S ~ phosphoglyceric acid + $DPNH_2$

III. enzyme-S ~ phosphoglyceric acid + orthophosphate ⇌
enzyme-SH + 1, 3-diphosphoglyceric acid

This mechanism, which strongly resembles that proposed for the oxidation of the α-keto acids, has been suggested in the same form by RACKER
(1951) for still another enzyme system, glyoxalase. This catalyses the
oxidation of methylglyoxal to pyruvate, and RACKER believes that an
enzyme-S-pyruvyl complex occurs as an intermediate in this reaction.

At this time, KRIMSKY and RACKER (KRIMSKY and RACKER 1952, RACKER and
KRIMSKY 1952) were studying the mechanism of the inhibition by Fe^{++} ions
of glycolysis in brain homogenates. They found that this inhibition could
be removed by addition of glutathione and were able to identify the glutathione effect as a re-activation of the phosphoglyceraldehyde dehydrogenase. Since glutathione was greatly superior to all other SH compounds
tested, it occurred to the authors that it might, in fact, be the prosthetic
group of the enzyme. In a very elegant series of experiments, in some of
which they employed isotopically labelled enzyme and glutathione, they
demonstrated that this was indeed the case.

This finding made it possible for the authors to present a detailed scheme of
the different stages in the dehydrogenase reaction. This is based on the following
evidence: (a) the isolation of the acyl-enzyme complex, (b) the formation of acyl-
glutathione in the presence of added glutathione and the absence of phosphate,
(c) the phosphorylation of added acyl-glutathione by the enzyme, (d) the
identification of glutathione as the prosthetic group of the enzyme, (e) the
fact that, if the oxidative properties of the enzyme are blocked by inhibiting
the prosthetic group with iodoacetate, the phosphorylation can proceed if acetyl-
glutathione is added, (f) the observation that the equilibrium of the aldolase
reaction can be displaced by addition of glutathione, which shows that the
3-phosphoglyceraldehyde formed in this reaction couples spontaneously with the
SH group of the glutathione.

On the basis of the data, the authors propose the scheme given in
Fig. 19 A for the reaction.

This scheme has been supplemented by the authors with evidence
regarding the manner in which DPN is involved in the system. This
evidence they have derived from the shift in the absorption spectrum of
DPN which occurs when it is bound to the enzyme. The complete reaction
mechanism, according to KRIMSKY and RACKER, is shown in Fig. 19 B.

It is assumed here that DPN is bound to the SH group of the prosthetic
group of the enzyme in a manner greatly resembling that proposed by
KAPLAN for the linkage between DPN and orthophosphate. The substrate
splits this bond oxidatively, i.e. the DPN is reduced, while the oxidized
substrate takes its place on the enzyme, being coupled to the latter by an
energy-rich S~acyl bond. The acyl group is transferred to phosphate

at a second active centre on the enzyme molecule. The prosthetic group of the enzyme is thus liberated. In the absence of phosphate, added SH compounds can also act as acyl-acceptors. It should be added that the enzyme catalyses yet a further reaction, namely the transfer of the phosphate group of the acyl compound to ADP.

It is a matter of future investigation whether this mechanism applies also to other oxidative systems. Nevertheless, the resemblance between this and the mechanism proposed for the oxidations of pyruvate and α-ketoglutarate is so striking (see Fig. 18, 1 *A–B*) that in would perhaps be more fruitful to point out the differences between the schemes than to re-emphasize the analogies. One difference is that the prosthetic group of the α-keto acid oxidases is apparently not glutathione but possibly LTPP. A further difference is that the acyl group is transferred in the phosphoglyceraldehyde to a mobile SH-bearing acceptor, *i.e.* CoA, before the phosphorylation occurs. A third point of dissimilarity lies in the phosphorylation itself, which here results either in acyl phosphate or phosphoryl-CoA, according as whether the enzyme derives from bacteria or from animal tissue, while in the triose phosphate dehydrogenase reaction acyl phosphate is invariably formed. KRIMSKY and RACKER believe that in the latter case the phosphorylation is catalysed by a single enzyme, while in the former case several different catalytic units are involved.

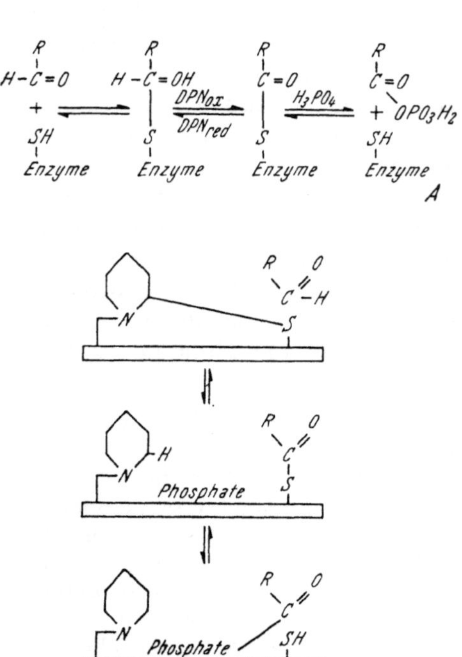

Fig. 19. The mechanism of phosphorylation coupled to the oxidation of 3-phosphoglyceraldehyde. (According to RACKER and KRIMSKY 1952.)

d) The DPNH₂-FAD system and the S-energy concept

If an attempt is now made to draw a parallel between KRIMSKY and RACKER's mechanism outlined above and the hypothesis put forward by KAPLAN, presented on p. 48, to explain the appearance of phosphate-bound energy in association with the electron transport from DPN to FAD, it is difficult not to feel that the demonstration of a phosphorylated intermediate at the aldehyde stage may also present difficulty in this latter case. If, however, the existence and identity of a SH-bearing prosthetic group should be demonstrated at some future date, this could without difficulty be incorporated in the scheme in the place of phosphate. The initial stage in KAPLAN's reaction mechanism (see Fig. 14) would then be:

$$
\begin{array}{ccccc}
CH & CONH_2 & & CH & CONH_2 \\
HC & C & & HC & C \\
HC & CH & + \text{HS-Enzyme} \rightarrow & HC & C - \text{S-Enzyme} \\
& N & & & NH \quad H \\
& R & & & R
\end{array}
$$

In this case also, phosphorylation would then occur as a secondary reaction (cf. Fig. 18, 3 *A*—*B*).

II. Energy-Transferring Processes

A. Conversion of Thiol-Bound Energy into Phosphate-Bound Energy

It has been commonly believed that the formation of ATP on the liberation of oxidative energy would become understandable as soon as the mechanism of coupling between oxidation and phosphorylation was elucidated. For many years, in fact, a number of reactions have been known where energy-rich phosphate groups are exchanged reversibly between ATP and other known energy-rich phosphate compounds, such as 1.3-diphosphoglyceric acid, phosphopyruvate and phosphocreatine. Thus, as long as it was thought that the primary energy-rich bond included a phosphate group, the problem of the synthesis of ATP was limited to the search for a "primary ester", which could give rise to ATP by a transphosphorylation reaction analogous with those mentioned above.

LIPMANN's discovery of acetyl phosphate was a promising contribution in this sense, for it could enter into reversible transphosphorylation with ADP:

$$\text{acetyl phosphate} + \text{ADP} \rightleftharpoons \text{acetate} + \text{ATP}$$

However, as has been mentioned at an earlier stage, it was shown that acetyl phosphate was not involved in LIPMANN's liver system:

$$\text{acetate} + \text{sulphanilamide} \xrightarrow{\text{ATP} + \text{CoA}} \text{acetylsulphanilamide}$$

It was therefore supposed that ATP activated acetate to a form not identical with acetyl phosphate, which could be transferred to sulphanilamide through the mediation of CoA. The same system could also acetylate hydroxylamine by a non-enzymic coupling of the latter with the "active" acetate. SOODAK and LIPMANN (1948), moreover, found another liver system which, in the presence of ATP and CoA, catalysed the condensation of two molecules of acetate to one molecule of acetoacetate.

A thorough investigation, including competition experiments (NOVELLI, GREGORY, FLYNN and SCHMETZ 1951) and fractionation of the different acetylating systems (CHOU, NOVELLI, STADTMAN and LIPMANN 1950) yielded definite evidence that "active" acetate was common to all these systems. It was suggested that it was, in fact, acetyl-CoA.

Further support for this was obtained (Novelli and Lipmann 1950) in experiments with an extract of *Escherichia coli,* which catalysed the condensation of oxaloacetate and acetate to citrate in the presence of ATP. Here it was found that acetyl phosphate not only replaced acetate and ATP, but even increased the rate of reaction. On the basis of this finding, the activation of acetate by acetyl phosphate could be represented in the following manner:

$$\text{acetyl phosphate} + \text{CoA} \rightarrow \text{acetyl-CoA} + \text{phosphate}$$

If this bacterial enzyme, named phosphotransacetylase and later purified by Stadtman, Novelli and Lipmann (1951), was mixed with the liver enzymes previously employed, acetylation could be brought about in the absence of ATP (Novelli, Gregory, Flynn and Schmetz 1951). It remained, however: (a) to isolate CoA, (b) to explain the activation of acetate by ATP in animal tissues, where acetyl phosphate could not be an intermediate, and (c) to demonstrate that acetyl-CoA actually contained a bond of an energy level equivalent to that of an energy-rich phosphate bond.

These problems were solved by the brilliant work of Lynen and co-workers (Lynen and Reichert 1951, Lynen, Reichert and Rueff 1951). From yeast extract these workers succeeded in isolating pure acetyl-CoA in which they demonstrated the nature of the S-acyl linkage. By substitution of the isolated acetyl-CoA in systems where the ATP-acetate system or acetyl phosphate had formerly been employed as energy donors, they succeeded in showing that the S-acyl bond occupied the same energy level as the energy-rich phosphate bond. This finding led the Lynen group to believe that energy bound in acetyl-CoA was converted into phosphate-bound energy in animal tissues, where transacetylase does not occur, by a phosphorolytic cleavage of acetyl-CoA. The phosphoryl-CoA thus formed could react with ADP to give ATP:

$$\text{acetyl-CoA} + \text{phosphate} \rightleftharpoons \text{phosphoryl-CoA} + \text{acetate}$$
$$\text{phosphoryl-CoA} + \text{ADP} \rightleftharpoons \text{CoA} + \text{ATP}$$

This hypothesis has subsequently acquired a more general validity, despite the fact that certain minor modifications have recently been found necessary for the special case of acetate.

Sanadi and Littlefield (1951), using a purified α-ketoglutaric oxidase, have demonstrated the occurrence of succinyl-CoA as a product of oxidation. Kaufman (1951) found a protein fraction which catalyzed the phosphorylation of ADP to ATP in the presence of succinyl-CoA. He concluded that the following reactions occurred:

$$\text{succinyl-CoA} + \text{phosphate} \rightleftharpoons \text{phosphoryl-CoA} + \text{succinate}$$
$$\text{phosphoryl-CoA} + \text{ADP} \rightleftharpoons \text{CoA} + \text{ATP}$$

While this reaction mechanism has been confirmed in different quarters (unpublished communications, International Biochemical Congress, Paris,

1952), there has been difficulty in establishing the validity of the acetyl transport mechanism proposed by Lynen and co-workers. The Lipmann school (Lipmann, Jones, Black and Flynn 1952, Lipmann, Jones and Black 1952) isolated the ATP-acetate system from yeast and pigeon liver. Their aim was to study, in the presence of hydroxylamine as acetyl acceptor, the reaction products in this system which, in accordance with the suggestions of Lynen and co-workers (see above), should catalyse the overall reaction:

$$ATP + acetate + hydroxylamine \xrightarrow{CoA} ADP + phosphate + hydroxamic\ acid$$

They found, however, that a liberation of orthophosphate in amounts equivalent to the hydroxamic acid formed did not occur. When, indeed, fluoride was added to the system, the formation of orthophosphate could be completely eliminated and could hence be attributed to the presence of phosphatase as a contaminant.

The subsequent investigation, which was based on a very elegant analytical technique, made it clear that pyrophosphate, and not orthophosphate, partook in the reaction. Lipmann therefore modified the Lynen mechanism to read:

$$acetyl\text{-}CoA + pyrophosphate \rightleftharpoons pyrophosphoryl\text{-}CoA + acetate,$$
$$pyrophosphoryl\text{-}CoA + AMP \rightleftharpoons CoA + ATP$$

These reactions have been confirmed with pyrophosphate, acetyl-CoA and AMP as reactants. Green (paper read at International Biochemical Congress, Paris, 1952) points out, however, that pyrophosphoryl-CoA is not accumulated in the system in the absence of adenylic acid. He suggests that the two component reactions are catalysed by the same enzyme and that pyrophosphoryl-CoA exists only at the enzyme level.

By way of summary, it may be said that the reversible reactions involving the acyl-CoA intermediates and ATP have to a large extent been elucidated. A further link in the transfer system is constituted by the transfer of the acyl group between the primary enzyme complex and CoA.

This was originally formulated by Schweet, Fuld, Cheslock and Paul (1951) as follows (see also p. 52):

$$acyl \sim enzyme + CoA \rightleftharpoons acyl \sim CoA + enzyme$$

The observation of the rôle of α-lipoic acid in this reaction and the discovery that this material occurs bound to cocarboxylase have provided evidence that LTPP is the SH-bearing prosthetic group of the oxidase. The reaction may thus be formulated (Reed and de Busk 1952 b):

$$acyl \sim LTPP + CoA \rightleftharpoons acyl \sim CoA + LTPP$$

The accompanying scheme (Fig. 20) provides a summary of the series of coupled reactions dealt with in the preceding pages, which have been shown to occur between primary reaction products and ATP.

A hypothetical transfer mechanism between pyridone-SH and ATP is also included. This latter is purely speculative. As has been pointed out above, there is considerable reason for believing that the primary product in this case also contains no phosphate. Whether the energy-rich S-pyridone linkage is directly split by phosphate, or whether the energy is initially conserved in a mobile pyridone carrier of the SH type (in analogy with CoA) is still unknown. Similarly, it cannot yet be decided whether the phosphorolytic cleavage results in a pyridone phosphate or an SH-phosphate.

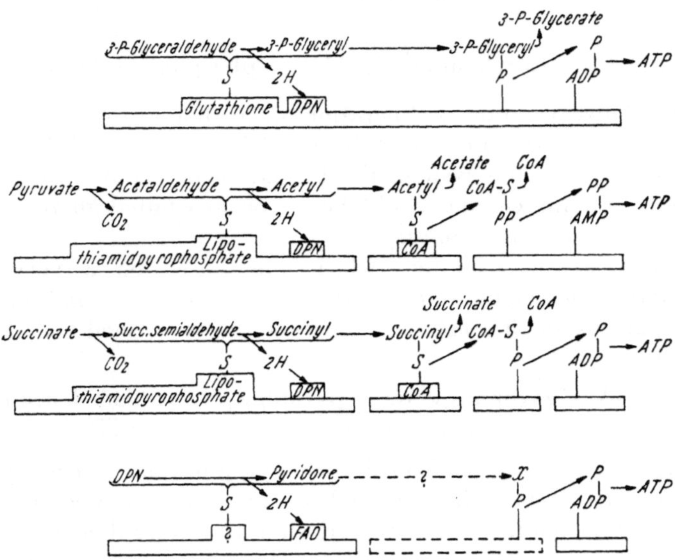

Fig. 20. Liberation and transfer of energy during oxidation of a) 3-phosphoglyceraldehyde; b) pyruvate; c) α-ketoglutarate, and d) DPN (hypothetical).
The basement lines indicate catalytic units.

B. Interrelationship of Energy-Transferring Processes

The above scheme may also be regarded in a different manner (Fig. 21). The various primary products may be conceived as being in reversible communication with the central ATP by way of radially arranged reaction chains. The intermediates on the "spokes" may serve as transmitters of energy for different activation reactions (see further p. 70). In this concept it would clearly be possible to find a reversible path of communication over ATP between any two energy-rich intermediates.

It will readily be realized that such an arrangement ensures a great flexibility of the energy-mediating system. Moreover, it provides the system with a "balancing device", capable of maintaining an appropriate concentration relationship of the different energy-transmitting units, *i. e.* members of the adenylic acid system and derivatives of SH-coenzymes, with regard to the fluctuations which may occur in the supply of substrate or the extent of energy utilization. The present section deals with some aspects of this balancing device.

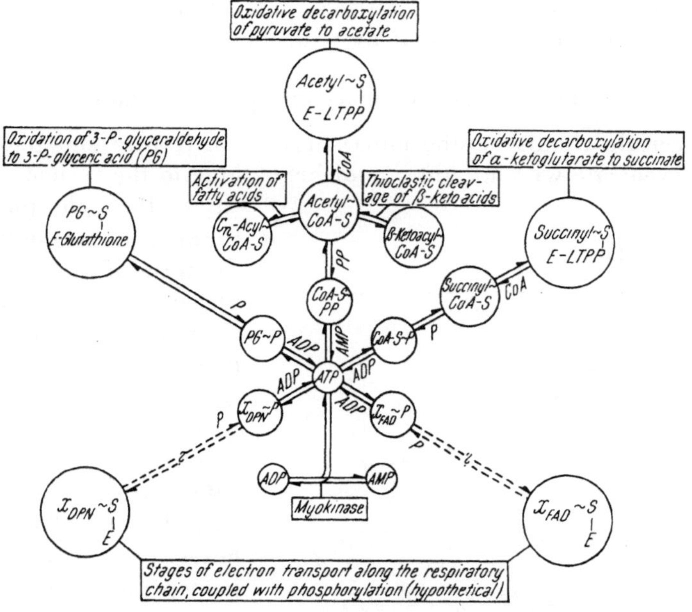

Fig. 21. Reversible transfer of energy between the primarily formed high-energy bonds and ATP. The rectangular labels indicate the levels of energy generation.

1. The myokinase reaction

The myokinase reaction:

$$2\,ADP \rightleftharpoons ATP + AMP$$

was discovered in 1943 by Colowick and Kalckar. The enzyme, as isolated from skeletal muscle, proved to be the most stable one known hitherto, with a remarkable resistance against heat, acid and alkali. The presence of myokinase in mitochondria was detected, in 1951, independently by Barkulis and Lehninger and by Kielley and Kielley. The particulate enzyme lacks the above stability and resembles the liver myokinase, purified and studied in the meantime by Kotelnikova (1950). Mitochondrial myokinase is sensitive to fluoride, as has been shown by Barkulis and Lehninger, and confirmed recently by Siekevitz and Potter (1953).

According to Barkulis and Lehninger, whose view was later shared by Slater and Holton (1952), the catalytic capacity of the mitochondrial myokinase is low as compared with the maximal rate of oxidative phosphorylation attainable in the system. Lindberg and Ernster (1952 a), on the other hand, found that addition of ADP, labelled with P^{32} in the labile phosphate group, to respiring mitochondria supplemented with glucose and hexokinase, resulted in a very rapid incorporation of the isotope into the glucose-6-phosphate formed. Moreover, the incorporation was not inhibited even by relatively high concentrations of fluoride. In view of these findings the question arises whether there is a difference in myokinase activity between "working" and "resting" mitochondria. The query seems in some degree justified by its analogy with the problem of ATPase.

2. The synthesis of inorganic pyrophosphate

Under certain conditions the mitochondria can respire and phosphorylate at an even rate without addition of hexokinase to the system. In such cases the phosphorylated product is accumulated in the form of inorganic pyrophosphate. Salt - prepared mitochondria of kidney and liver (cyclophorase) readily accumulate pyrophosphate under these conditions (Kornberg and Lindberg 1948, Cross, Taggart, Covo and Green 1949), while heart mitochondria show a perceptibly smaller tendency to do this (Lindberg and Ernster 1951). If mitochondria are prepared in isotonic sucrose, their power to accumulate pyrophosphate is diminished (cf. Kielley and Kielley 1951), while it is completely lost (unpublished observation) if they are prepared by the method of Slater and Cleland (1952) under such conditions that their surface properties, such as their selective permeability (Cleland 1952), are preserved in the highest possible degree.

Even in those mitochondria which cannot accumulate pyrophosphate, a certain concentration of this material nevertheless appears during aerobic phos-

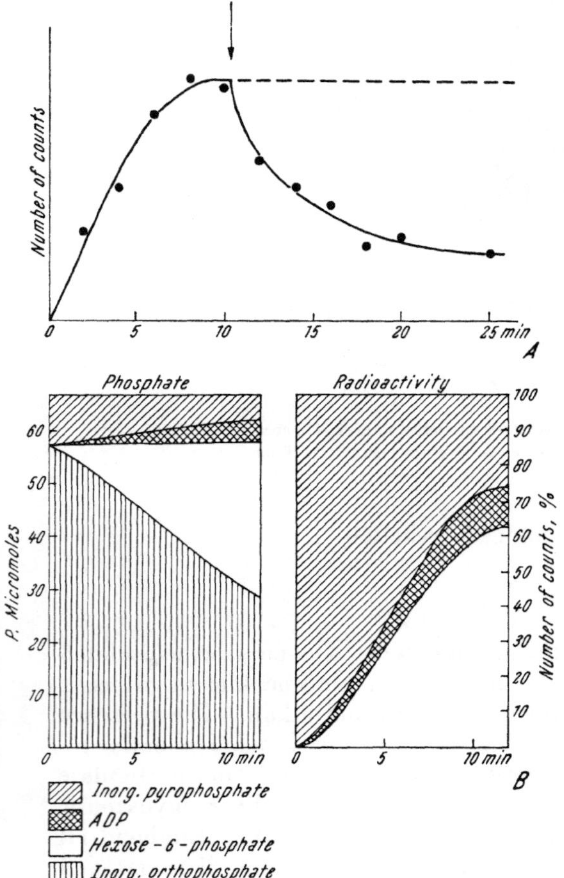

Fig. 22. Experiments demonstrating the metabolism of inorganic pyrophosphate during oxidative phosphorylation in isolated rabbit heart mitochondria supplemented with yeast hexokinase.

A) Pyrophosphate is formed from a source of labelled orthophosphate during oxidative phosphorylation. When pyrophosphate had attained a steady state level (dotted line), unlabelled orthophosphate was added (arrow). The solid curve shows the decrease of the radioactivity of the pyrophosphate pool. B) Oxidative phosphorylation in the presence of labelled inorganic pyrophosphate. The radioactivity (figure on the right) appears in the hexose-6-phosphate formed, without having passed the orthophosphate stage.

(For experimental details see Lindberg and Ernster 1952.)

phorylation in the preseence of an excess of hexokinase. This concentration remains constant as long as phosphorylation in the system continues at a constant rate (LINDBERG and ERNSTER 1951). It has been established that the pyrophosphate, in this system, is not an accumulation product, but an active intermediate: in an experiment where isotopically labelled pyrophosphate was formed the radioactivity of the pyrophosphate gradually disappeared when unlabelled orthophosphate was added. In the same investigation it was also found that the rate of turnover of the pyrophosphate correspondend to about 5% of the phosphorylation rate of the mitochondria. It could be demonstrated, furthermore, that added labelled pyrophosphate was transferred to glucose-6-phosphate by way of ATP (Fig. 22).

On the basis of these findings a hypothesis was formulated according to which adenine nucleotides act in two ways as acceptors of high energy phosphate. The main way, responsible for about 95% of the total phosphorylating capacity of the seystem, consists of a transfer of an orthophosphate group from a primary ester to ADP, yielding ATP. A second, quantitatively less significant pathway involves a pyrophosphorylation of AMP. It was suggested that this reaction was identical with the dinucleotidase reaction of KORNBERG (1950):

$$\text{dinucleotide} + \text{pyrophosphate} \rightleftharpoons \text{mononucleotide} + \text{ATP},$$

the only reaction at that time known to involve inorganic pyrophosphate, and the occurence of which in mitochondria had already been deduced by other workers (GREEN 1951 b).

Later on, however, HOGEBOOM and SCHNEIDER (1952) demonstrated that the dinucleotidase was not present in mitochondria, but occured exclusively in the cell nucleus. At the same time, the discovery by LIPMANN and his co-workers (LIPMANN, JONES, BLACK and FLYNN 1952) of the pyrophosphorylysis of acetyl-CoA suggesteed a new interpretation of the metabolic pathway of pyrophosphate:

$$\text{ATP} \underset{\text{AMP}}{\overset{\text{CoA}}{\rightleftharpoons}} \text{pyrophosphoryl-CoA} \underset{\text{acetyl-CoA}}{\overset{\text{acetate}}{\rightleftharpoons}} \text{pyrophosphate}$$

Disregarding the great significance this reaction may play in the activation of fatty acids by ATP, it clearly presents a mechanism by which ATP can be formed by a pyrophosphorylation of AMP. The old problem whether or not ADP is the only acceptor of high energy phosphate in a mitochondrial system thus becomes a question as to whether or not inorganic pyrophosphate can be formed in such a system without interaction of adenine nucleotides.

The above reaction mechanism cannot, on the other hand, explain a continous accumulation of pyrophosphate with a simultaneous maintenance of a constant rate of respiration, as was observed with liver and kidney mitochondria under the conditions indicated above. The accumulation of pyrophosphate would in this case involve a progressive displacement of the equilibrium in the pyrophosphoryl-CoA reaction, with the consequence that free acetate would accumulate at the expense of critic acid formation. We therefore suggest that pyrophosphate accumulation in these structurally damaged systems is a result of a cleavage of pyrophosphoryl-CoA or phosphoryl-CoA according to one of the reactions:

(1) 2 phosphoryl-CoA → 2 CoA + pyrophosphate

(2) pyrophosphoryl-CoA → CoA + pyrophosphate

Such a mechanism would, in fact, not influence the continuity of the Krebs cycle.

3. The synthesis of acetoacetate

The above hypothesis receives some support from considerations of the problem of acetoacetate synthesis, a matter which has been studied by many workers in connection with the mechanism of fatty acid oxidation (for review, see Barker 1952 and Green 1951 b).

As has been mentioned earlier (p. 35), the oxidation of a fatty acid is preceeded by an activation reaction in which the acid is bound to CoA. The activated acid (for the sake of simplicity, octanoic acid will be taken as an example) is oxidized in two stages to the corresponding β-keto compound:

$$\text{octanoyl-CoA} \xrightarrow{-4\,\text{H}} \beta\text{-ketooctanoyl-CoA}$$

which is then split reversibly by CoA:

$$\text{CoA} + \beta\text{-ketooctanoyl-CoA} \rightleftharpoons \text{acetyl-CoA} + \text{hexanoyl-CoA}$$

Further oxidation and splitting gives, in an analogous manner, another molecule of acetyl-CoA and a molecule of butyryl-CoA. (If the starting material were a fatty acid with an *odd* number of carbon atoms, propionyl-CoA would be formed, the further oxidation of which is considered on p. 36.) The butyryl-CoA is oxidized in two stages to acetoacetyl-CoA.

In muscle and in kidney the degradation of acetoacetyl-CoA proceeds in a manner analogous with the preceding stages and it is split with the aid of CoA to give two molecules of acetyl-CoA:

$$\text{acetoacetyl-CoA} + \text{CoA} \rightleftharpoons 2 \text{ acetyl-CoA}$$

In these organs, therefore, the even-numbered fatty acids are degraded quantitatively to acetyl-CoA units. The condition for complete oxidation to carbon dioxide and water is thus, as in the oxidation of pyruvate, that a condensing partner is available. If such a substance is not present, pyruvate and fatty acids cannot be oxidized in these systems.

Another state of affairs prevails in liver. The acetyl-CoA units formed by β-oxidation can be subjected, here too, to further oxidation, provided that, as in the case of pyruvate, a condensing partner is present. If this, on the other hand, is not so, two acetyl-CoA units condense irreversibly with the formation of acetoacetate and two molecules of CoA (Soodak and Lipmann 1948). At the same time, acetoacetyl-CoA cannot undergo reversible degradation in the presence of CoA to two molecules of acetyl-CoA, as was the case in muscle and in kidney, but is split irreversibly to CoA and acetoacetate. Liver thus has two irreversible mechanisms for the synthesis of acetoacetate:

(1) 2 acetyl-CoA $\rightarrow$ 2 CoA + acetoacetate
(2) acetoacetyl-CoA $\rightarrow$ CoA + acetoacetate

but it lacks the reversible one for splitting acetoacetyl-CoA, which implies that it cannot oxidize acetoacetate.

On the basis of these observations, we should like to formulate an analogy between the metabolic patterns of acetoacetate and pyrophosphate: muscle and liver differ in their pattern of acetoacetate metabolism in the same way as do intact and damaged mitochondria *in vitro* in their metabolism of pyrophosphate. Liver mitochondria appear to be much more susceptible to such damage than are heart mitochondria.

4. The mode of action of 2,4-dinitrophenol

It appears at the present time, as discussed above, that phosphorylation is not primarily associated with the mechanism for the conservation of the energy liberated in cell oxidations. None of the coupling reactions at present known involves the participation of phosphate groups. If this proved to be true for those coupling reactions still unknown, it would imply that phosphate takes part only in reversible transfer reactions as an energy-carrier, in the same manner as the adenylic acid system. This possibility constitutes a serious warning against the interpretation of turnover rates of ATP and other energy-rich phosphate compounds, as measured in experiments with radioactive phosphorus, as indices of the metabolic activity of tissues.

A realization of the probability that phosphate is not involved in the actual coupling mechanism, puts the effects of 2,4-dinitrophenol (DNP) and related inhibitors in a new light. While it has previously been considered that these substances are uncoupling agents, it now appears necessary to reject this view and to conclude that they in fact exert an effect upon the transport system.

The idea of employing DNP as an inhibitor derives from the observation of Warburg (1926) (cf. also Lynen and Koenigsberger 1951) that a number of substances, to which DNP was later shown to belong, had the power of inhibiting the Pasteur effect. In 1948, Loomis and Lipmann found that DNP blocked phosphorylation without inhibiting respiration. Since the respiration required to be accompanied by phosphorylation, they interpreted their finding as an indication that DNP could replace phosphate.

This conclusion was later contradicted by Teply (1949). This author found that DNP furthered the dephosphorylation of the labile phosphate esters occurring in cyclophorase preparations ("gel phosphate"). cf. Green, Atchley, Nordmann and Teply (1949) and Albaum (1949). By the splitting of these substances, orthophosphate could continually be made available for the phosphorylation processes associated with respiration. Teply showed considerable foresight here, since this was the first observation of the effect of DNP in inducing phosphatase-like processes in mitochondria.

Hunter (1951) extended Teply's observations and showed that DNP also stimulated the ATPase in mitochondria, an effect previously known in intact muscle (Ronzoni and Ehrenfast 1936) and muscle extract (cf. Hunter 1951). Hunter and Spector (1951) found, moreover, that the uncoupling effect of DNP was limited to phosphorylations taking place *below* the substrate level. This had earlier been supposed to be the case by Lynen and Koenigsberger (1951) in view of the results of their studies in the effect of DNP on yeast.

A summary of the views of Teply and of Hunter on the DNP effect is given in Fig. 23. Both these authors believe the effect to consist in the stimulation of a phosphatase, which attacks "the primary ester" (Teply) or *one* of "the primary esters" (Hunter). The question arises, however, whether a pre-existing enzyme is activated or whether the phosphatase in question may be formed under the influence of the DNP.

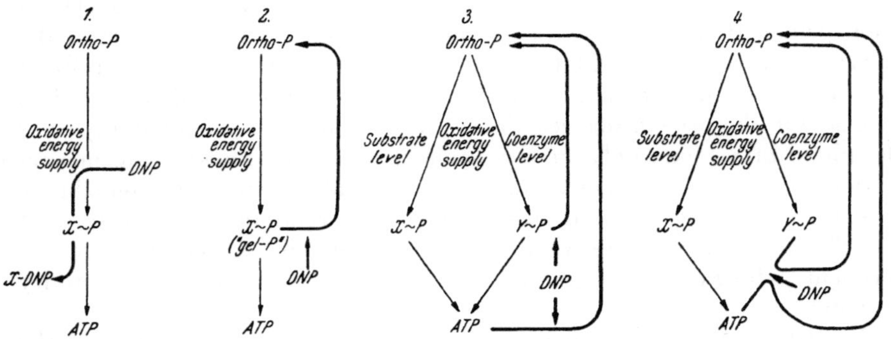

Fig. 23. Hypotheses concerning the mode of action of 2, 4-dinitrophenol (DNP).
1) Loomis and Lipmann (1948). 2) Teply (1949). 3) Hunter (1951). 4) Present work.

Some support is given to the latter alternative by the fact that a previously non-existent ATPase activity in mitochondria is developed by treatment with DNP. Kielley and Kielley (1951) have shown, and this was later confirmed by Potter and Recknagel (1951), that mitochondria subjected to hypotonic conditions or to ageing display an increasing ATPase activity. Mitochondria prepared in ionic media are also rich in ATPase. In general, the ATPase activity is the weaker, the more fully the mitochondrial structure is retained.

It is evident that DNP develops the ATPase activity in parallel with the phosphatase of Teply and Hunter which attacks "primary esters". It has been proposed by Hunter (1951), at the suggestion of Pinchot, that the phenomenon does not consist in the activation of two pre-existing phosphatases, but rather in the cleavage into two phosphatases of a transphosphorylase acting between the "primary ester" and ATP.

In fact, if, after the treatment with DNP, the transphosphorylase activity persisted in addition to that of the activated phosphatases, phosphorylation could be made to proceed in the presence of an adequate supplement of hexokinase. Salt mitochondria, which show a partial DNP effect, in that they exert an ATPase activity, but in which transphosphorylation is not completely eliminated, have strong phosphorylating activity in the presence of hexokinase. This indicates that hexokinase can compete with the increased phosphatase effect, provided that a part of the transphosphorylase persists.

Concentrations of DNP which uncouple phosphorylation have a stimulating effect upon oxidation *in vivo* (cf. PEISS and FIELD 1948). This stimulation is absent with isolated mitochondria that are provided with optimal concentrations of phosphate and adenylic acid. In this manner DNP can be employed to demonstrate that phosphate or a phosphate acceptor constitutes a limiting factor in a biological system (cf. LYNEN and KOENIGSBERGER 1951, CLELAND, unpublished). JUDAH and WILLIAMS-ASHMAN (1951) have shown that adenylic acid is essential for maximum oxidation in isolated mitochondria even in the presence of DNP. These experiments required a preparation of mitochondria completely free from adenine nucleotides. It might, however, be thought that such mitochondria would not be capable of phosphorylation even at the substrate level, so that the necessity of AMP, in spite of the presence of DNP, would be understandable.

In contrast to the case of the intermediates of the Krebs cycle, the oxidation of fatty acids is blocked by DNP (CROSS, TAGGART, COVO and GREEN 1949). This phenomenon has been considered to be associated with the property of DNP of uncoupling phosphorylation and the conclusion is drawn that the activation of the fatty acids, which is a prerequisite for their oxidation, involves a phosphorylated intermediate (GREEN 1952). It must, however, be pointed out that the phosphorylation and pyrophosphorylation of CoA, and the transfer of the phosphate groups to ADP, are not blocked by DNP. It is known, on the other hand, that the activated forms of the fatty acids are their compounds with CoA. The question thus arises whether it is justifiable to deduce a causal relationship between the inhibiting effects of DNP on fatty acid oxidation and on the phosphorylations which lie below the substrate level. As has been stated above, it seems probable that the blocking of oxidative phosphorylation consists in a cleavage of a transferring enzyme catalysing the transfer of phosphate from the primary phosphate compound (say pyridone phosphate) to ATP. The fragments from this cleavage would then acquire the properties of phosphatases, which would split the primary compound and ATP respectively. If transferring enzymes could be converted by such cleavage into phosphatases or other hydrolytic enzymes, this would explain why phosphorylation is unessential to certain oxidases in the mitochondrial system, such as succinic oxidase (CROSS, TAGGART, COVO and GREEN 1949) or β-hydroxybutyric oxidase (LEHNINGER 1951). It might well be conceived that the transferring enzymes in these cases are so labile that they can spontaneously undergo such fission in the absence of phosphate or ADP. Other transferring enzymes, such as the transphosphorylases in the DPN-FAD region, might be more stable and be degraded only in the presence of DNP. This category of higher stability would then include the transferring enzymes catalysing the activation of the fatty acids. A third category of enzymes, the transphosphorylases at the substrate level, would be so stable as to be unaffected by DNP.

As has previously been mentioned, effects similar to that of DNP, though in a smaller degree, can be brought about by methods which may be regarded as affecting the mitochondrial surface (treatment with hypotonic media, detergents, etc.) (cf. Hunter 1951). It will be an interesting matter for future study to determine the relationship between the stability of the individual enzymes and their position within the mitochondria. In the light of our present limited knowledge, it is tempting to imagine that the oxidase systems are the more labile, the greater their proximity to the mitochondrial surface. In such an event, the succinic oxidase system would have a position nearest the surface[2], while the pyruvic and a-ketoglutaric oxidases would be less peripherally situated.

III. Energy-Utilizing Processes

In the two preceding sections we saw that the mitochondria are responsible for the liberation of almost all the energy placed at the disposal of the cell in the form of food, and for a transformation of this energy into forms suitable for syntheses and the performance of work. We found also, however, that in their energy-transforming activity, the mitochondria are dependent upon the energy-utilizing processes. This mutual dependence of the endergonic and exergonic processes frequently rendered it difficult to define a boundary, in the metabolic respect, between those energy-utilizing reactions which take place within the mitochondria and those in which non-mitochondrial enzymes are active.

In order that the synthetic reaction

$$A + B \xrightarrow{\text{enzyme, energy}} AB$$

shall be possible, not only are the substances A and B and a specific enzyme necessary, but also the availability of energy in a suitable form. However this energy be provided to the system, it is necessary in the first instance that it be transferred either to A or B, so that this component may couple with the other at the expense of its energy content. This process may be called "activation".

Let us take the hexokinase reaction as an example, where A represents phosphate and B glucose. In order to activate the phosphate we may proceed in various manners. If we couple the system to mitochondria, which oxidize a substrate in the presence of adenylic acid, the phosphate is activated by conversion to ATP by way of a series of transfer reactions such as has previously been described. In combination in ATP the phosphate can finally be coupled to glucose with the aid of hexokinase. Another means of activating the phosphate would be to introduce energy into the system in the form of acetyl-CoA. In this case respiring mitochondria would not be required, although the presence of the mitochondrial enzyme complement catalysing the reversible transfer of energy between acetyl-CoA and ATP would still be necessary. In both these cases the transferring

---

[2] While this work was in proof. Cleland and Slater (1953) presented experimental data which enabled them to draw the conclusion that the succinic oxidase system is associated with the mitochondrial membrane.

enzymes of the mitochondrial system would be called upon. These, on the other hand, as has previously been discussed in detail, could only carry out the process of activation if the energy-utilizing mechanism, in this case the hexokinase reaction, were functioning. A third way of bringing about the hexokinase reaction would be by direct addition of ATP, in which case, obviously, no other systems would be required.

It follows from this example that a treatment of the synthetic reactions related to the mitochondria must embrace not only those actually located in the mitochondria, but also the extramitochondrial reactions dependent upon the mitochondrial activation processes, where the dependence is, as has been emphasized, reciprocal.

A further reason may be given for the consideration of these extra-mitochondrial reactions, namely that observations have recently been made which indicate that the tapping of energy from active mitochondria by means of added enzyme systems possibly takes place at the mitochondrial surface. It appears, at least where heart mitochondria are concerned, that ATP and other energy-rich compounds can penetrate the surface only with difficulty (unpublished observations). In order to explain the action of added hexokinase, it has been suggested that the hexokinase reaction takes place at the mitochondrial surface (ERNSTER and LINDBERG 1952). This had previously been supposed to be the case by HUNTER (1951). It has recently been reported by CRANE and SOLS (1953) that the glucokinase in animal tissues is bound to particles to a large extent and that the bound enzyme displays properties differing from those of the enzyme in solution.

The most urgent problem today in connection with synthetic processes, appears to be the elucidation of the activation mechanism and the identification of the active form of the substrate. In most isolated systems hitherto studied ATP can serve as a source of energy. The activation may consist in a one-stage reaction, provided that the transfer of energy from ATP takes place in the form of a transphosphorylation. In such a case, phosphorylated intermediates should be present. In several instances, however, such as the syntheses of glutathione, hippuric acid and citrulline, it has been shown that low-molecular co-factors are involved. This indicates that at least two transfer reactions take place, one between ATP and the co-factor and one between the co-factor and the substrate.

In a systematic treatment of the synthetic reactions related to the mitochondria, it appears suitable, in view of the above considerations, to classify the reactions in accordance with the mode of activation whereby energy is made available.

A. ATP as Primary Source of Energy

1. Kinase reactions

This group of enzymes catalyzes the phosphorylation of alcohol groups with the aid of ATP, which functions here as active phosphate. It is customary to name the enzymes after the alcohols they phosphorylate, with the suffix "kinase".

The earliest to be discovered was hexokinase, which in animal tissue, as distinct from yeast, is a collective term for a number of enzymes, each being specific for its own hexose. It is thus possible to distinguish between glucokinase (for review, see Colowick 1951 b), and fructokinase, which has recently been purified and shown to phosphorylate fructose in the 1-position (Leuthardt and Testa 1951). Phosphofructokinase phosphorylates fructose-6-phosphate to fructose-1,6-diphosphate. This enzyme has been purified by Racker (1947). Kidney tissue has been shown to phosphorylate glycerol. The enzyme has been demonstrated by Kalckar (1939) and studied by Lindberg (1951). It has been called glycerokinase. A number of additional enzymes have recently been added to this list, which catalyse the phosphorylation of nucleosides and other aromatic compounds of alcohol character. To these belong adenosinekinase (Kornberg and Pricer 1951, Caputto 1951), riboflavinkinase (Kearney and Englard 1951), thiaminekinase (Leuthardt and Nielsen 1952) and pyridoxalkinase (Hurvitz 1952).

The intracellular location of these enzymes is at present unclear. The thiamine-phosphorylating enzyme has, however, been found to occur in the ground cytoplasm (Leuthardt and Nielsen 1952). The same applies to fructokinase (Leuthardt and Testa 1951). In the case of glucokinase, Hers, Berthet, Berthet and de Duve (1951) have found that it is principally located in the soluble fraction. Crane and Sols (1953), however, as has already been mentioned, found that the enzyme was partly present in particulate elements. They found also that the rate of the reaction fell off, with accumulation of glucose-6-phosphate. If the latter was converted to fructose-1,6-diphosphate with the aid of added enzymes, the reaction could be made to proceed at a constant rate.

2. Muscle contraction

The ATPase activity of myosin can be included among the exergonic reactions in which ATP serves as a primary source of energy. Here no synthetic process is involved, but rather a conversion of chemical into mechanical energy. We must therefore modify the term "activation" in this case to signify the chemical reaction by which the actomyosin complex is enabled to contract. It is hardly possible at present to decide whether the hydrolysis of ATP releases contraction or whether the energy of ATP is first transferred to the protein and is only liberated at this stage. Certain reports in the literature (Szorenyi and Chepinoga 1946, Gelotte 1951, Snellman and Gelotte 1951) suggest that the ATPase mechanism of myosin does not involve a direct liberation of inorganic orthophosphate, but the transfer of a phosphate group to an acceptor bound to the actomyosin. In such a case ATP would not represent the final stage in the transport of energy. A clarification of the rôle of the protein SH groups may help towards a solution of the problem in the near future.

The location of the mitochondria in relation to the contractile elements of muscle, a matter studied by Watanabe and Williams (1951) and Harman and Feigelson (1952 c), will be dealt with more fully in the next chapter (p. 100). These studies provide a good opportunity of observing, on the morphological level, the nature of the relationship between the energy-generating and energy-utilizing mechanisms.

3. The formation of active methionine

A new type of reaction, in which ATP acts as "active adenosine" rather than "active phosphate", has recently been described by Cantoni (1951, 1952). It consists of a coupling of adenosine with methionine and takes place in association with the splitting of phosphate from ATP. The reaction is catalysed by an enzyme isolated from liver and serves to produce the form of methionine which methylates nicotinamide and guanidoacetic acid with the aid of the appropriate transmethylases. The chemical constitution of this methionine derivative is reported by Cantoni to correspond to an adenosine-5-S-methionine ester, containing a tertiary sulphonium linkage. There is no information at present concerning the intracellular distribution of the enzyme.

B. CoA Compounds as Primary Source of Energy

A review of the reactions, steadily increasing in number. in which CoA is known to partake. has recently been made by Lipmann, Jones and Black (1952). A common feature of all these processes is that they involve the coupling of a carboxylic acid to an acceptor. The activation of the acid consists in the coupling of its carbonyl group to CoA. This activation makes it possible for acyl groups to condense with different acyl acceptors. amines. alcohols, etc. in accordance with the specific character of a given acylating enzyme.

In the case of acetate the peculiar phenomenon is found that the acetyl-CoA can not only act as an acetyl donor, but can also condense by way of its methyl group. This is the case in the condensation with oxaloacetate to give citrate and with acetate to give acetoacetate. A satisfactory chemical interpretation of this behaviour, called "head and tail acetylation" by Lipmann, has not yet been made.

A formal analogy between these processes and the kinase reactions considered above, may be made in the following manner:

ATP ("active phosphate") + acceptor → phosphorylated acceptor + ADP
acyl-CoA ("active acyl") + acceptor → acylated acceptor + CoA

This parallel has been emphasized by Lipmann and co-workers in their review. where they suggest that the acyl-transferring enzymes should be given the common name of acetokinases. In order that the analogy should be made even more clear. it would offer much advantage to replace the prefix "aceto" by the name of the acceptor. This would make it possible to differentiate between the various acetylating enzymes as in the case of the phosphate-transferring enzymes.

The treatment of the individual reactions will cover the synthesis of citrate, acetylated amines and acetylcholine. together with certain recent findings concerning tho rôle of succinyl-CoA in the synthesis of pyrrole ring systems and the formation of phosphatidic acids from CoA-activated higher fatty acids. The synthesis of acetoacetate, which really comes under the same heading, has already been considered in detail (p. 66). In conclusion, some detoxication reactions, specific to the liver, where an activation by CoA is involved, will be discussed.

The distribution of all these enzymes is far from being elucidated. The syntheses of citrate and of acetoacetate, together with the formation of hippuric and p-aminohippuric acids, are mitochondrial processes. The formation of acetylcholine and the acetylation of sulphanilamide and glucosamine are catalysed by non-mitochondrial enzymes.

1. The synthesis of citric acid

The problem of citrate synthesis has been so intimately concerned with that of the nature of the active two-carbon body, that an historical account here would merely involve repetition. The discovery of CoA made it possible for the Ochoa school, who had long been concerned with the citrate problem, to study in detail the mechanism and kinetics of the condensation reaction (Stern and Ochoa 1951, Ochoa, Stern and Schneider 1951, Stern, Shapiro, Stadtman and Ochoa 1951, Korkes, del Campillo, Gunsalus and Ochoa 1951, Stern, Ochoa and Lynen 1952). The crystalline enzyme, isolated from pig heart, catalyses the reaction:

$$\text{acetyl-CoA} + \text{oxaloacetate} \rightarrow \text{citrate} + \text{CoA}$$

The reaction involves an energy loss of about 8000 cal.

The synthesis of citrate is clearly a mitochondrial process. This implies that mitochondria are involved in the maintenance of the physiological citrate level and thus that the mechanism responsible for the regulation of that level must exert, directly or indirectly, a regulatory action on the activity of the mitochondria.

It has been observed in several quarters that under certain conditions added citrate cannot be oxidized, or only very feebly, by isolated mitochondria (Plaut and Plaut 1952, Lindberg, Ljunggren, Ernster and Révész 1953). Some other authors, principally those who have isolated mitochondria in a salt medium, find that citrate is oxidized as rapidly as other intermediates of the Krebs cycle (cf. Green 1952). In order to explain the failure of citrate to undergo oxidation in the experiments of the first-named group of workers, the possibility has been considered that the isocitric dehydrogenase is a limiting factor. This enzyme, as mentioned at an earlier stage, is present in the mitochondria only in 12% of its total quantity.

That this circumstance cannot be held accountable, however, has been conclusively shown by Plaut and Plaut (1952) on the basis of the following observations: a) $C^{14}H_3COOH$ oxidized in the presence of α-ketoglutarate carrier gives radioactive α-ketoglutarate; b) in the presence of citrate, the same substrate gives radioactive citrate on oxidation. The carbon dioxide formed is also radioactive, but its specific activity shows that the citrate formed and that added are not in exchange equilibrium. Also Peters (1952), in his work on the inhibition of the aconitase reaction by fluorocitrate, came to the conclusion that citrate can enter the mitochondrial body only slowly, if at all.

The permeability of mitochondria, on the other hand, is dependent on the concentration of Ca^{++}, as has recently been shown by CLELAND (1952). If Ca^{++} is present, most ions pass the membrane freely. If, however, the calcium is bound in complex formation with added versene, a selective permeability is observed. Since citrate has a chemical effect similar to that of versene, it may be imagined that, by binding calcium, it prevents its own passage into the mitochondria.

The biological importance of such a regulation mechanism cannot as yet be assessed. The dramatic effect brought about by an increase in the citrate content of the blood by immobilization of Ca^{++} (PETERS 1952, MARTIUS 1952) may well be due to a disturbance of the permeability-regulating mechanism.

2. Acetylation of amines

The acetylation of amines *in vitro* was observed at an early date by the LIPMANN school and the history of these investigations has been related in earlier chapters. All these reactions involve the acetylation of an amine through the mediation of CoA. The individual enzymes can be isolated from an extract of pig liver by fractionation with acetone. A general scheme for the preparations has recently been given by CHOU and LIPMANN (1952). The amines for which acetylation enzymes have been found in the liver extract are sulphanilamide, *p*-aminobenzoic acid, sulphathiazol and sulphathiazine (LIPMANN 1945), histamine (MILLICAN, ROSENTHAL and TABOR 1949), glucosamine and chrondosamine (CHOU and SOODAK 1952).

3. The acetylation of choline

The formation of acetylcholine is one of the earliest and most intensively studied of the acetylation reactions. The enzyme concerned has aroused great interest in view of its important part in the mechanism of nerve excitation. A comprehensive review of the subject has recently been published by NACHMANSOHN and WILSON (1951). The enzyme. which catalyses the reaction:

$$\text{acetyl-CoA} + \text{choline} \rightarrow \text{acetylcholine} + \text{CoA}$$

has been obtained from *Octopus* ganglia by fractionation of the aqueous extract of an acetone powder (KOREY, DE BRAGANZA and NACHMANSOHN 1951. KORKES, DEL CAMPILLO, KOREY, STERN, NACHMANSOHN and OCHOA 1952). The above overall reaction, as a typical kinase reaction, corresponds to an energy loss of about 8000 cal. This has special interest in the present case. since NACHMANSOHN (1951) believes that this amount of energy is expended in binding acetylcholine to a membrane protein.

4. The synthesis of protoporphyrin

The only synthetic reaction at present known which involves succinyl-CoA is the protoporphyrin synthesis (Radin, Rittenberg and Shemin 1950 a, b, Shemin and Wittenberg 1951, Shemin and Kumin 1952). The experiments were carried out on duck blood corpuscles and haemolysate, with isotopically labelled succinate, acetate and glycine. By a remarkable combination of technique and ingenuity, the authors came to the conclusion that the precursor of the pyrrole nucleus in the citric acid cycle is a metabolite intermediate between α-ketoglutarate and succinate, in all probability succinyl-CoA. They suggest that two molecules of this substance combine with one molecule of glycine to form a pyrrole derivative in the following manner:

$$
\begin{array}{ccccc}
 & COOH & & & COOH \\
 & | & & & | \\
COOH & CH_2 & & COOH & CH_2 \\
| & | & & | & | \\
CH_2 & CH_2 & & CH_2 & CH_2 \\
| & | & & | & | \\
CH_2 + & COR & \rightarrow & CH_2 \cdots COR \\
| & & & & \\
COR & CH_2 - COOH & & COR & CH_2 - COOH \\
 & N & & & N \\
 & H_2 & & & H_2
\end{array}
$$

Four molecules of this derivative are then coupled through the carboxyl carbon atom of the glycine to give a molecule of protoporphyrin.

5. The synthesis of phosphatides

According to a recent communication by Kornberg and Pricer (1952), compounds of CoA with higher fatty acids (C_{12}–C_{18}) can couple with α-glycerophosphate with the formation of mono- and di-esters, the so-called phosphatidic acids. The enzyme has been isolated from rat liver. The reaction mechanism, studied by addition of C^{14}-stearic acid, P^{32}-α-glycerophosphate, ATP and CoA, is described as follows:

$$\text{stearate} + \text{CoA} \xrightarrow{\text{ATP}} \text{stearyl-CoA}$$

stearyl-CoA + α-glycerophosphate $\rightarrow$ monostearylphosphatidic acid + CoA

stearyl-CoA + monostearylphosphatidic acid $\rightarrow$ distearylphosphatidic acid + CoA

Kennedy (1952) has reported that liver mitochondria prepared in a sucrose medium, when carrying out oxidative phosphorylation in the presence of P^{32}, catalyse a renewal of lipids. He thus provides confirmation of earlier reports by FRIEDKIN and LEHNINGER (1949 a). The lipid turnover was manifested by an incorporation of P^{32}, and of C^{14}-myristic acid, and it could be stimulated by addition of glycerol. These facts indicate that the mechanism responsible for the renewal of phospholipids in the presence of glycerol, orthophosphate and fatty acids, possibly including the above reactions formulated by KORNBERG, is present in the mitochondria.

An incorporation *in vitro* of C^{14}-acetate into fatty acids of lipids has recently been reported by BRADY and GURIN (1952). The enzyme system catalysing this process was obtained by extracting an acetone powder of pigeon liver mitochondria with microsome-free supernatant. The reaction took place both anaerobically and aerobically and was stimulated in the latter case by Mg^{++}, DPN, cytochrome *c* and especially by citrate. Oddly enough, acetate proved to be a better precursor of fatty acids than acetyl-CoA. No reaction was obtained with mitochondrial extract or supernatant alone.

It hence appears that mitochondria contain the enzymes necessary for the total renewal of phosphatides with the exception of that of their fatty acid components. Even fatty acids can, as we know, be catabolized by native or acetone-treated mitochondria (see p. 49), while their synthesis seems to require the participation of certain components from the ground cytoplasm.

6. The synthesis of hippuric and *p*-aminohippuric acids

Processes which appear to be unambiguously associated with liver and kidney mitochondria are the syntheses of hippuric and *p*-aminohippuric acids. The enzyme catalysing the latter has been located and the requirements for the system have been studied by COHEN and McGILVERY (1947). The functions of CoA had not been revealed at the time of this investigation and it was believed that a phosphorylated intermediate was involved in the reaction. As a source of energy, ATP could be employed. For the maintenance of a continuous reaction, however, it was necessary that an oxidative phosphorylation were carried on by the system. *N*-phosphoglycine could not replace ATP and glycine.

The results have been confirmed by the studies of KIELLEY and SCHNEIDER (1950) on the same system. These authors obtained data regarding the intracellular distribution of the enzyme, thus confirming the earlier report that this was almost exclusively present in the mitochondria.

More recent work by CHANTRENNE (1951) has shown that the synthesis of hippuric acid from glycine and benzoic acid in the presence of ATP and acetone powder of rat liver extract requires the presence of CoA. Benzoyl phosphate cannot act in the system. From these results, it seems that the rôle of ATP is to couple either benzoic acid or glycine with CoA. The same coupling can obviously be brought about with the aid of energy deriving from respiratory processes.

It remains to determine whether it is benzoic acid or glycine which is coupled to CoA. Cohen (1951) reasons that if *benzoate* couples to give an active benzoyl-CoA, this implies a generalization of the rôle of CoA in carboxyl activations; if, on the other hand, *glycine* couples to CoA over its carboxyl group, it remains to explain how this activation can affect the reactivity of the amino group. A general function of CoA in this respect follows from Green's (1951 b) reasoning with regard to the mechanism of CoA-catalysed reactions.

Leuthardt and Nielsen (1951) have found, however (cf. also Green 1951 b), that while benzoic acid in the hippuric acid synthesizing system can be replaced by a large variety of both aliphatic and cyclic carboxylic acids, glycine can only be replaced by substances which may give rise to glycine in the same system. When glycine is thus substituted, an increased synthesis of hippuric acid is observed.

C. Reactions Involving Unknown Energy-Rich Intermediates

1. The conjugation of phenol. "Active sulphate"

The mechanism of conjugation of phenol has been studied *in vitro* by de Meio and Tkatz (1950) and by Bernstein and McGilvery (1952). The reaction involves a conjugation of phenol with sulphuric acid or glucuronic acid. In liver homogenate the two processes occur side by side, although the formation of phenyl sulphate is predominant by a factor of 10 to 15. The conjugation requires aerobic conditions and is inhibited by 2, 4-dinitrophenol. De Meio and Tkatz have established that mitochondria and supernatant together are necessary for the formation of phenyl sulphate.

With regard to the activation mechanism, Bernstein and McGilvery have arrived at the interesting and somewhat unexpected conclusion (cf. Green 1951 b) that the sulphate and not the phenol is activated, while CoA does not participate in the process. The evidence for active sulphate is based on an elegant experiment where the enzyme preparation, incubated with sulphate in the absence of phenol, was found to form a substance which could then be conjugated with phenol; if, however, the phenol was incubated in the absence of sulphate, no such active compound was obtained. The proof of the occurrence of active sulphate is of great interest in relation to the mechanism of synthesis of other sulphuric acid esters of biological importance.

2. Carbon dioxide fixation

Utter and Wood found in 1946 that the incorporation of $C^{14}O_2$ into the β-carboxyl group of oxaloacetate catalysed by liver preparations was stimulated by ATP. The enzyme, which was subsequently partially purified by Vennesland, Evans and Altman (1947), showed, in contrast to the carboxylases of microorganisms, a definite requirement of ATP. In the consideration of the function of ATP, the following possibilities have been considered (UTTER and WOOD 1951, UTTER 1951):

a) ATP synthesizes a co-factor essential for the CO_2 fixation reaction;

b) ATP inhibits a reaction interfering with the fixation;

c) ATP activates the enzyme protein;

d) ATP partakes directly in the fixation reaction.

In their experimental investigation, however, the same authors concluded that none of these alternatives could represent the whole truth, although there were certain indications in favour of one or the other. It was thus possible to show that the synthesis of a co-factor with the aid of ATP could be excluded. At the same time, however, it was found that glutathione had a strong stimulating effect upon the reaction if it was

Table 6. *Stimulation of fixation of $C^{14}O_2$ in oxaloacetate by ATP and glutathione.* (From UTTER 1951.)

Additions	β-COOH of oxaloacetate counts/min/mg C
None	16
ATP	654
Glutathione	102
ATP + glutathione . .	2168

Pigeon liver extract, dialyzed for 17 hrs. ATP. 2 μ moles; glutathione. 2.5 μ moles. 50,000 counts of C^{14} in $NaHC^{14}O_3$ in 100 μ moles. Total volume 2.0 ml, incubated for 5 min. at 38° C.

added simultaneously with ATP (Table 6). This effect was especially marked with enzyme that had been subjected to prolonged dialysis.

Another interesting finding was that there was no apparent equivalence between the amount of ATP utilized and that of carbon dioxide fixed; the carbon dioxide fixation exceeded many times the stoichiometric equivalent of ATP, even if the latter was calculated in terms of energy-rich phosphate bonds.

The crude dialysed enzyme was then submitted to further purification, during which it was divided into two fractions. One of these catalysed almost exclusively the incorporation of carbon dioxide and not of pyruvate, while both fractions together caused the incorporation of carbon dioxide and pyruvate in almost equal proportions (Table 7).

From these experiments, it seems that pyruvate is not a primary product in the decarboxylation of oxaloacetate. It therefore seems reasonable to suggest that pyruvate occurs in an activated form. Since the experiments showed conclusively that no phosphorylated intermediate was involved. but that glutathione greatly stimulated the reaction, it is tempting to imagine that the active intermediate is a sulphhydryl-bound pyruvyl radical. The occurrence of such a compound in the reaction catalysed by the yeast enzyme glyoxalase, has, indeed, been proposed by RACKER (1951). In such an event, the fixation reaction would have the following form:

(1) pyruvate + ATP + RSH $\rightleftharpoons$ pyruvyl $\sim$ SR + ADP + orthophosphate

(2) pyruvyl $\sim$ SR + CO_2 $\rightarrow$ oxaloacetate + RSH

There are, however, two difficulties in the acceptance of this scheme. It does not explain why carbon dioxide can be fixed in amounts greater than correspond

Table 7. *Fixation of $NaHC^{14}O_3$ and $CH_3.CO.C^{14}OONa$ by fractions of pigeon liver extracts.*

(From UTTER 1951.)

	Residual oxaloacetate		Ratio β-COOH to α-COOH
Pigeon liver fraction	β-COOH, representative of incorporation of CO_2	α-COOH, representative of incorporation of pyruvate	
	cts/min/mg C	cts/min/mg C	
Dialyzed crude	628	312	2.01
51—59% am. sulf. fraction . . .	559	45	12.4
51—59% + 0—51% am. sulf. fractions	859	665	1.29
51—59% + 59—100% am. sulf. fractions	624	45	13.9

All vessels contained 100 μM of $NaHC^{14}O_3$ containing 50,000 counts, and 100 μM of $CH_3.CO.C^{14}OONa$ containing 27,500 counts, and 2 μM of ATP, 120 μM oxaloacetate, 4 μM $MnCl_2$ in a total volume of 0.2 ml. Incubated for 5 min. at 38° C.

to the amount of ATP utilized; nor does it indicate why fixation of carbon dioxide is not obtained with CO_2 and pyruvate in the absence of oxaloacetate.

With regard to the latter point, it may be asked whether it is the incorporation of carbon dioxide into, or its removal from, oxaloacetate that requires ATP. UTTER maintains that, since the fixation of carbon dioxide is an endergonic reaction, it is logical to suppose that the part of ATP is to provide energy for the process. On the other hand, it has recently been reported by GREEN, BEINERT, FULD, GOLDMAN, PAUL and SARKAR (1953) that in enzyme preparations derived from cyclophorase and capable of oxidizing all other Krebs cycle intermediates, oxaloacetate was not oxidized in the absence of ATP. If oxaloacetate were decarboxylated by the system of GREEN et al.—and there is good reason to suppose that the necessary enzyme was present—the oxidation of oxaloacetate would doubtless have proceeded in a manner analogous to that of pyruvate. It therefore appears possible that not only the incorporation, but also the removal of carbon dioxide requires ATP. GREEN and co-workers come to the interesting conclusion that the aerobic oxidation of malate to oxaloacetate does not take place in their system unless ATP is present.

With these circumstances in mind, we would like to suggest the possibility (cf. GREEN 1951 c) of an activation process in connection with the malic dehydrogenase reaction. The process might be represented thus:

(3) malate + RSH $\rightleftharpoons$ malate-SR

(4) malate-SR + pyridine   nucleotide $\rightarrow$ *enol*oxaloacetate $\sim$ SR + red.   pyridine nucleotide.

This mechanism would imply that oxaloacetate does not occur as a free intermediate and would hence explain the necessity of ATP for both the oxidation and the decarboxylation of free oxaloacetate.

Since reaction (4) is not present in UTTER and WOOD's carboxylation system. the rôle of ATP would here consist rather in the activation of oxaloacetate than in that of pyruvate as has been suggested above (reaction 1). The reaction mechanism might thus be supposed to be:

(5) oxaloacetate + RSH + ATP $\rightleftharpoons$ *enol*oxaloacetate $\sim$ SR + ADP + orthophosphate

(6) *enol*oxaloacetate $\sim$ SR $\rightleftharpoons$ pyruvyl $\sim$ SR + CO_2

Since both these component reactions are reversible, there is no longer a necessity for stoichiometric equivalence between ATP consumed and carbon dioxide incorporated. Apart from an initial activation of oxaloacetate, the process is no longer an endergonic reaction, but a reversible transfer at the high-energy level. The reaction rate hence depends exclusively on the turnover number of the enzyme, after equilibrium has been reached between the left- and right-hand sides in the overall reaction. The part of ATP would thus be to initiate the reaction by activation of oxaloacetate and to partake in the overall equilibrium together with orthophosphate and ADP. UTTER himself actually suggests the possibility that ATP "aids in the equilibration of the added labelled CO_2 and CO_2 arising from decarboxylation", which is in harmony with the above hypothesis.

The assumption of a transfer at the high-energy level, instead of an endergonic mechanism, for the oxaloacetate carboxylase reaction provides a possible means of escape from the confusion arising from the fact that OCHOA's malic enzyme, in contrast to the oxaloacetic carboxylase, catalyses the incorporation of carbon dioxide in the absence of ATP, if a source of $TPNH_2$ is provided (OCHOA 1951). It might be conceived that the carboxylation in this reaction takes place before the oxidation and thus proceeds at the low-energy level:

(7) malate-SR $\rightleftharpoons$ methylglyoxal-SR + CO_2,

while the oxidation, and hence the elevation of the system to the high-energy level, takes place only in the next stage:

(8) methylglyoxal-SR + pyridine nucleotide $\rightarrow$ pyruvyl $\sim$ SR + red. pyridine nucleotide

If this mechanism is to be confirmed. it will be necessary to take up again the old question of the occurrence of methylglyoxal in animal systems. just as has recently been done in the case of acetaldehyde (p. 51).

3. The synthesis of glutamic acid

The synthesis of glutamic acid from α-ketoglutarate and ammonia has not yet been studied with the isolated enzyme. Particulate preparations catalyse the synthesis under anaerobic conditions, when the following dismutation takes place:

$$2\,\alpha\text{-ketoglutarate} + NH_3 \rightarrow \text{glutamate} + \text{succinate} + CO_2 + H_2O$$

This synthesis was studied many years ago by Krebs and Cohen (1939), who suggested that the formation of glutamate took place in two stages:

$$\alpha\text{-ketoglutarate} + NH_3 \rightarrow \text{iminoglutarate} + H_2O$$
$$\text{iminoglutarate} + 2\,H \rightarrow \text{glutamate}$$

More recently, Hunter and Hixon (1949) and Hunter (1951) have investigated this system in rat liver and found that ammonia and phosphate are necessary for the dismutation. The necessity for ammonia indicates that the mechanism put forward by Krebs is correct and that the incorporation of ammonia is essential for the reduction of α-ketoglutarate.

On the other hand, it was reported as early as 1929 by Hahn and Harmann (cf. Martius 1952) that in the anaerobic "fermentation" of citric acid in muscle and in liver homogenate, α-hydroxyglutaric acid accumulated.

Our knowledge is still insufficient to permit a decision between the following alternatives: a) whether the formation of glutamic acid from α-ketoglutaric acid and ammonia involves first a reversible incorporation of ammonia at the low-energy level to give iminoglutarate, followed by a reductive elevation of the energy level to glutamate, or b) whether there is an initial reductive elevation of the energy level to hydroxyglutarate and then a reversible introduction of ammonia. In the fixation of carbon dioxide in oxaloacetate, discussed in the foregoing section, both types of reaction seem to occur.

4. Synthesis of γ-glutamyl derivatives

a) Glutamine

The synthesis of glutamine was first described in 1935 by Krebs in slices of brain, kidney and retina, and by Leuthardt and Bujard (1947) in liver homogenate. The last-named workers were the first to show that the system required the presence of ATP.

Subsequent investigations by Speck (1949) and Elliot (1951) have shown that the reaction takes place as follows:

$$\text{glutamate} + ATP + NH_3 \rightleftharpoons \text{glutamine} + ADP + \text{orthophosphate}$$

A strict stoichiometric ratio of orthophosphate liberated to glutamine formed has been established, while a limitation of the reaction by deficient provision of ammonia also limits the liberation of orthophosphate. This suggests that the occurrence of glutamyl phosphate, which has been envisaged as an obvious intermediate, is not conceivable, since in this case the phosphorylation of glutamic acid would proceed in spite of the lack of ammonia. No co-factor has been demonstrated, although it seems probable that some such substance participates (cf. Cohen 1951).

The rôle of a co-factor would be to couple to the γ-carboxyl group of the glutamate, utilizing for this purpose the energy of ATP, whereupon the active glutamate would react with ammonia to give glutamine. The latter is a clearly exergonic reaction, since the amide bond, like the peptide bond, represents an energy of 3—4000 cal.

The intracellular site of the glutamine synthesis has been intensively studied by FREI and LEUTHARDT (1949). They found that a combination of mitochondria and the supernatant fraction was necessary for the synthesis. Since it is known that glutamine can be metabolized in mitochondria (LEUTHARDT, MÜLLER and NIELSEN 1949) and since, furthermore, there is reason to suppose that any activation of glutamic acid would be associated with the mitochondria, it is difficult at present to interpret this finding.

b) Glutathione

The synthesis of glutathione has been studied by JOHNSTON and BLOCH (1951) in extracts of acetone-dried pigeon liver. Such extracts contain an enzyme which catalyses the synthesis of glutathione from glutamic acid, glycine and cysteine, and another enzyme which degrades the tripeptide hydrolytically. The two enzymes may be separated by high-speed centrifugation. Neither appears to be identical with the transpeptidase described by HANES, HIRD and ISHERWOOD (1950), which is said not to occur in liver.

The synthesizing enzyme requires ATP as a source of energy. The mechanism of activation is unknown and there is no clear evidence for phosphorylated intermediates. According to a report by SNOKE and ROTHMAN (1951), a heat-stable co-factor of unknown nature is essential to the synthesis. These authors have succeeded in separating the enzyme into two fractions, one of which catalyses the condensation of synthetic γ-glutamylcysteine and glycine, while both together bring about the synthesis of the tripeptide from the amino acids. In both cases the same co-factor is essential.

c) The synthesis of peptides

Considerable progress has recently been made by SIEKEVITZ (1952) in the study of the part played by mitochondria in protein synthesis. He has, by extending the earlier experiments by BORSOOK, DEASY, HAAGEN-SMIT, KEIGHLEY and LOWY (1950), demonstrated an incorporation of C^{14}-dl-alanine into the protein of rat liver microsomes in the presence of respiring and phosphorylating mitochondria (Fig. 24). No such uptake took place in the absence of mitochondria; nor did the alanine enter the mitochondrial protein in the absence of microsomes. The incorporation also failed to occur in the presence of dinitrophenol or hexokinase. If the mitochondria were incubated without microsomes in the presence of the materials necessary for respiration and phosphorylation, i.e. α-ketoglutarate, phosphate and adenylic acid, and the particles were spun down after a period of incubation, the supernatant contained a factor which, when added to microsomes alone, sufficed to bring about the incorporation of alanine. The factor concerned is not a protein. It can tolerate boiling for 7 minutes with N-HCl. According to the authors, it is neither "active alanine" nor ATP, but may be a substance derived from the latter.

7*

In an important publication, HANES, HIRD and ISHERWOOD (1950) report that enzyme preparations from kidney and pancreas, but not from liver, catalyse a transfer of the γ-glutamyl group of glutathione to other amino acids with the formation of γ-glutamyl peptides. The authors accordingly put forward the hypothesis that glutamine and glutathione, both having a γ-glutamyl function, induce transpeptidations with the γ-glutamyl group as the active principle.

They suggest that four types of transpeptidation may occur: a) the formation of γ-glutamyl peptides, b) the conversion of these into α-glutamyl peptides and c) and d) transpeptidations between two α- or two γ-peptides, respectively. According to the hypothesis all peptide synthesis can ultimately be referred to the synthesis of glutamine or glutathione (in plants, also that of asparagine) which, by the above processes, would be able to give, by reversible transpeptidation, any peptide whatever.

This basic hypothesis has now been formulated in a more elaborate form (for review, see WAELSCH 1952) according to which the exergonic phase of peptide synthesis in the formation of a γ-glutamylamide or -peptide. This "prototype" can then give rise to other peptides by transpeptidations in the following manner:

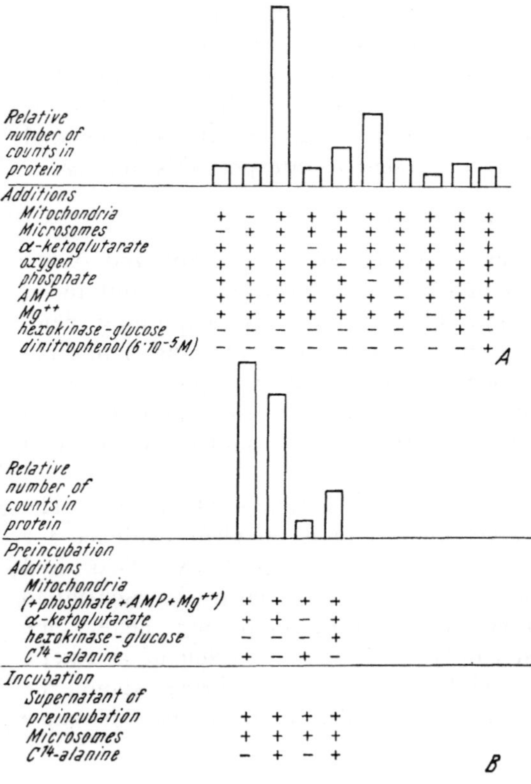

Fig. 24. Requirements for incorporation *in vitro* of C^{14}-alanine into proteins of particulate cell fractions.

A) Experiments involving a single incubation of particulate fractions under conditions varied as indicated in the figure. C^{14}-alanine added to each sample. B) Experiments involving a preincubation of the mitochondrial fraction, followed by an incubation of microsomes combined with the supernatant of the centrifuged preincubated suspension.

(After SIEKEVITZ 1952.)

(1) γ-glutamyl amino acid A $+\gamma$-glutamyl amino acid B $\rightarrow\gamma$-glutamyl amino acid A amino acid B $+$ glutamic acid

or

(2) γ-glutamyl amino acid A $+$ amino acid B $\rightarrow$ amino acid A amino acid B $+$ glutamic acid

The same principle would also be applicable to the prolongation of a peptide chain by single amino acids.

This formulation actually dispenses with the necessity of assuming a $\gamma \rightarrow \alpha$ conversion: the γ-glutamyl group is regarded rather as a catalyst of peptide formation. It should be pointed out that the "prototype" need not be glutamine, glutathione or asparagine: there may be γ-glutamyl compounds which are still unknown. The factor of SIEKEVITZ may be one of these.

A further discussion of protein synthesis is to be found on p. 108.

5. The synthesis of urea

Our knowledge of the synthesis of urea, which is based on the classical formulation by KREBS of the ornithine cycle, has been greatly augmented in recent years. The KREBS ornithine cycle involves three stages:

1. The coupling of a molecule of ammonia and a molecule of carbon dioxide to ornithine with the formation of citrulline

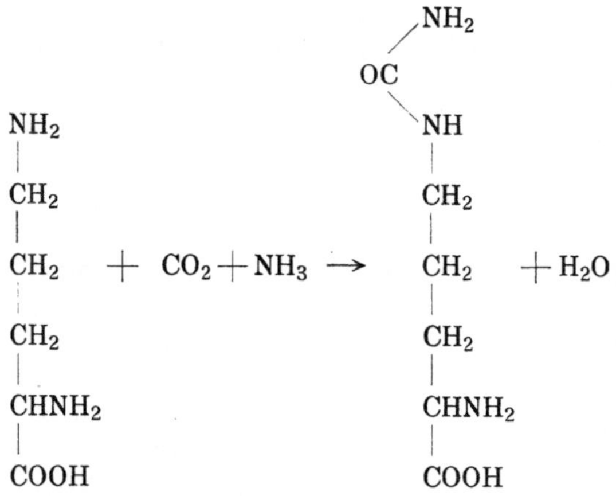

$$
\begin{array}{ccc}
& & OC{\nearrow NH_2 \atop \searrow NH} \\
NH_2 & & | \\
| & & CH_2 \\
CH_2 & & | \\
| & & CH_2 \\
CH_2 & + CO_2 + NH_3 \rightarrow & CH_2 & + H_2O \\
| & & | \\
CH_2 & & CH_2 \\
| & & | \\
CHNH_2 & & CHNH_2 \\
| & & | \\
COOH & & COOH
\end{array}
$$

2. An amination of citrulline with the formation of arginine. The amino group for this reaction is provided specifically by aspartic acid, as has been shown by BORSOOK and DUBNOFF (1941).

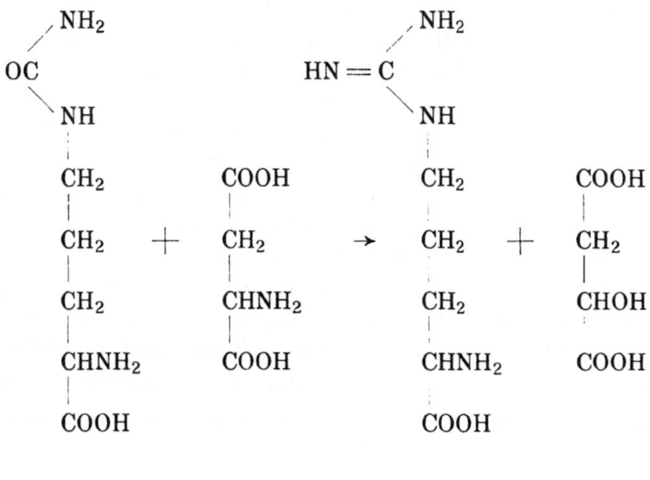

$$
\begin{array}{cccc}
OC{\nearrow NH_2 \atop \searrow NH} & & HN = C{\nearrow NH_2 \atop \searrow NH} & \\
| & COOH & | & COOH \\
CH_2 & | & CH_2 & | \\
| & CH_2 & | & CH_2 \\
CH_2 + CH_2 \rightarrow & CH_2 + CH_2 \\
| & | & | & | \\
CH_2 & CHNH_2 & CH_2 & CHOH \\
| & | & | & | \\
CHNH_2 & COOH & CHNH_2 & COOH \\
| & & | \\
COOH & & COOH
\end{array}
$$

3. A hydrolysis of arginine to ornithine and urea:

$$arginine + H_2O \longrightarrow ornithine + CO(NH_2)_2$$

It is now known that the first and second reactions each consist of at least three stages and that reaction (1) also requires a co-factor, α-carbamylglutamic acid. The formation of this substance and the processes involved in reaction (1) have been studied in especial detail by Cohen, Grisolia and co-workers (for review. see Grisolia 1951 and. as a later publication, also Grisolia and Cohen 1952). The mechanism of reaction (2) has been investigated by Ratner and co-workers (for review, see Ratner 1951). The relationship of the urea synthesis with the citric acid cycle and with the mitochondria has been elucidated, largely by Leuthardt and his co-workers (for review. see Leuthardt 1952).

The reaction mechanism in its more detailed form now appears to be as follows:

Glutamate fixes NH_3 and CO_2 with the aid of ATP, which results in the formation of α-carbamylglutamic acid:

$$
\begin{array}{l}
COOH \\
| \\
CH_2 \\
| \\
CH_2 \\
| \\
HOOC - CH \\
| \\
NH_2
\end{array}
\quad + CO_2 + NH_3 \xrightarrow{\text{ATP}} \quad
\begin{array}{l}
COOH \\
| \\
CH_2 \\
| \\
CH_2 \\
| \\
HOOC - CH \\
\quad\quad\quad | \\
\quad\quad\quad NH \\
OC\!\!< \\
\quad\quad\quad NH_2
\end{array}
\quad + H_2O
$$

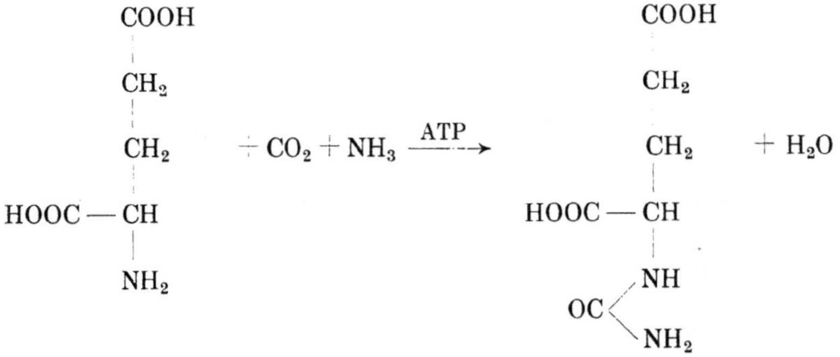

This reaction is catalysed by liver mitochondria. ATP is believed to provide energy for the fixation. although no phosphorylated intermediate has as yet been identified.

α-carbamylglutamic acid plays a catalytic rôle in the coupling of ammonia and carbon dioxide to ornithine. It has been shown conclusively that the carbamyl group of the α-carbamylglutamate does not partake in the reaction. The enzyme is bound to the liver mitochondria. The reaction is dependent upon the availability of ATP. which is consumed in amounts stoichiometrically equivalent to the citrulline formed.

The identity of an energy-rich intermediate has not been established. although there is a deal of evidence of its chemical nature. It is said to be a compound of the type phosphoryl-α-carbamylglutamate-NH_3-CO_2. The phosphate bond is extremely labile to acid and its presence in the molecule is essential for the formation of citrulline. It seems probable that the reaction takes place in two stages: a phosphorylation of α-carbamylglutamate, followed by a fixation of carbon dioxide and ammonia:

$$\alpha\text{-carbamylglutamate} + ATP \rightleftharpoons \alpha\text{-carbamylglutamate} \sim P + ADP$$

$$\alpha\text{-carbamylglutamate} \sim P + CO_2 + NH_3 \rightarrow \text{unknown intermediate}$$

Reaction (1) may thus be written:

ornithine + unknown intermediate → citrulline + α-carbamylglutamate

The amination of citrulline to arginine is a cytoplasmic reaction occurring in liver and kidney. The isolated enzyme requires aspartic acid as a source of ammonia, and also ATP. An intermediate condensation product of citrulline and aspartic acid has been detected and isolated. Reaction (2) may hence be represented in two stages:

$$
\begin{array}{l}
\mathrm{NH} \qquad\qquad \mathrm{COOH} \qquad\qquad \mathrm{NH} \qquad\quad \mathrm{COOH} \qquad\qquad \mathrm{NH} \qquad\qquad \mathrm{COOH}\\
\|\qquad\qquad\quad\ |\qquad\qquad\qquad\ \|\qquad\qquad\ |\qquad\qquad\qquad\ \|\qquad\qquad\qquad |\\
\mathrm{COH} \quad + \mathrm{H_2N-CH} \xrightarrow{\ \mathrm{ATP}\ } \mathrm{C-NH \cdot CH} \xrightarrow{\ \mathrm{H_2O}\ } \mathrm{C-NH} \ + \ \mathrm{CHOH}\\
|\qquad\qquad\qquad\ |\qquad\qquad\qquad\ |\qquad\quad\ |\qquad\qquad\qquad\ |\qquad\qquad\qquad |\\
\mathrm{NH} \qquad\qquad \mathrm{CH_2} \qquad\qquad\ \mathrm{NH} \qquad\ \mathrm{CH_2} \qquad\qquad\ \mathrm{NH} \qquad\qquad \mathrm{CH_2}\\
|\qquad\qquad\qquad\qquad\qquad\qquad |\qquad\quad\ |\qquad\qquad\qquad\ |\qquad\qquad\qquad |\\
\mathrm{CH_2} \qquad\qquad \mathrm{COOH} \qquad\quad \mathrm{CH_2} \quad\ \mathrm{COOH} \qquad\quad \mathrm{CH_2} \qquad\qquad \mathrm{COOH}\\
|\qquad\qquad\qquad\qquad\qquad\qquad |\qquad\qquad\qquad\qquad\ |\\
\mathrm{CH_2} \qquad\qquad\qquad\qquad\quad\ \mathrm{CH_2} \qquad\qquad\qquad\ \mathrm{CH_2}\\
|\qquad\qquad\qquad\qquad\qquad\qquad |\qquad\qquad\qquad\qquad\ |\\
\mathrm{CH_2} \qquad\qquad\qquad\qquad\quad\ \mathrm{CH_2} \qquad\qquad\qquad\ \mathrm{CH_2}\\
|\qquad\qquad\qquad\qquad\qquad\qquad |\qquad\qquad\qquad\qquad\ |\\
\mathrm{CHNH_2} \qquad\qquad\qquad\quad \mathrm{CHNH_2} \qquad\qquad\ \mathrm{CHNH_2}\\
|\qquad\qquad\qquad\qquad\qquad\qquad |\qquad\qquad\qquad\qquad\ |\\
\mathrm{COOH} \qquad\qquad\qquad\qquad\ \mathrm{COOH} \qquad\qquad\ \mathrm{COOH}
\end{array}
$$

The enzyme has been fractionated into two compounds. One of these is very labile and catalyses the condensation reaction; the other is stable, catalysing the hydrolysis of the condensation product to arginine and malate. Considerations of energy indicate that ATP must partake in the condensation reaction. There is experimental evidence that the labile condensing enzyme has two components, one being responsible for a reaction between ATP and either citrulline or aspartic acid, thus giving an active intermediate, the condensation of which is catalysed by the other enzymic component.

The requirement of aspartic acid by the isolated enzyme system is specific. If, however, the system is supplemented with mitochondria, all the amino acids related to aspartic acid by transamination reactions can also serve as sources of ammonia, provided that catalytic amounts of α-ketoglutaric acid are present. If the system is supplemented with mitochondria which perform oxidative phosphorylation in the presence of glutamic acid as substrate, glutamic acid can function as a source of both aspartic acid and phosphate-bound energy.

The enzyme arginase, which catalyses the hydrolysis of arginine with the formation of urea and the regeneration of ornithine, is found widely distributed in all animal organs. It does not appear to be confined intracellularly to any single cell constituent (SCHEIN and YOUNG 1952).

A general scheme of the reactions involved in urea synthesis, and their relation to the Krebs cycle, is found in Fig. 25.

6. The synthesis of *p*-amino-ornithuric acid

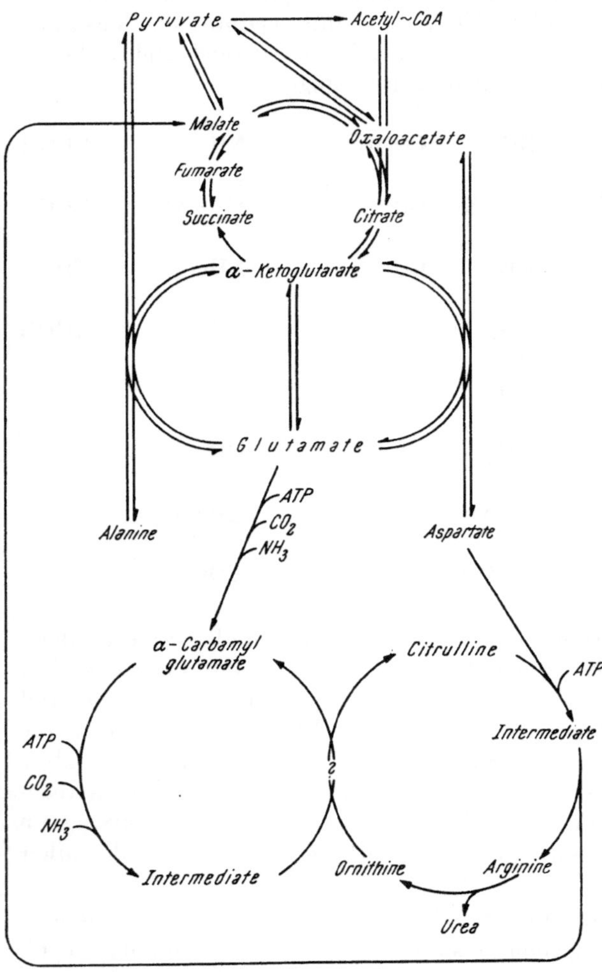

The synthesis of *p*-amino-*l*-ornithuric acid in chicken kidney mitochondria has been studied by McGilvery and Cohen (1950). The reaction involves a condensation of *p*-aminobenzoic acid with α-benzoyl-*l*-ornithine. ATP acts as energy donor, although no intermediates have as yet been found. Among the possibilities for such intermediates considered by the authors are the phosphorylated forms of the substrates, compounds of the substrates with ADP and complexes of the enzyme with phosphate or ADP. As a further alternative may perhaps be added that CoA or another SH-coenzyme may be involved, as is the case in the conjugation of *p*-amino-benzoic acid with glycine in mammalian liver.

Fig. 25. Urea synthesis and its relation to the Krebs cycle. (Partly after Ratner 1951.)

Physiology of Cytoplasmic Particles

A. Mitochondria

Morphological methods were for a long time the only means available for the study of cytoplasmic particles. They revealed particles of many functional types which were believed to be different varieties of mitochondria. It is, however, difficult to incorporate these in a general scheme.

It has long been envisaged, for example, that the mitochondria have a fundamental rôle in secretory processes, and investigations along these lines might well lead to a general understanding of the secretory mechanism of the living cell. Such studies would, however, contribute little to an understanding of the function of mitochondria in cells which do not secrete under physiological conditions.

By the same token, it may be tempting to attribute to the mitochondria absorptive and detoxicating functions in tissues specialized for these functions. It may well be, indeed, that studies of mitochondria will finally permit a general formulation of the mechanisms of absorption and detoxication. Here again, however, the basic function of the mitochondria will be made no clearer, and this function, as the two examples serve to illustrate, must provide the common basis for these two types of physiological activity.

We may name further examples: sarcosomes, kinetosomes, cortical granules and pigmented granules of various types display activities which are further branches on the common trunk. If the trunk is disregarded, there may still occasionally appear to be relations between the branches, but these are fortuitous. The bitter scepticism which is sometimes expressed by the cytologists (cf. *e. g.* RITCHIE and HAZELTINE 1953) certainly arises from desperation at this state of affairs, which is comparable to that in which an anatomist would find himself, were he to attempt to reconstruct a hydra from a cross section through the tentacles.

The single beacon in this obscurity, as has been very rightly emphasized by the cytologist (DE ROBERTIS, NOWINSKI and SAEZ 1948), is the study of the enzymic properties of isolated mitochondria. The important stages on the road which has led the enzymologist to the study of mitochondria are described in an earlier chapter. It may now be said, perhaps with a shade of wisdom after the event, that the enzyme chemist has taken the common mitochondrion, that which the cytologist would define as a particle stainable with Janus green, and attempted to draw the enzymic pattern which he sees there.

1. The metabolic spectrum of mitochondria

It appears to be a fundamental principle, the importance of which is still insufficiently realized, that the biological formation of a chemical bond with an energy content of some 3000 calories involves the breaking of an energy-rich bond of about 10,000 calories; it is, in other words, a strongly exergonic reaction. This is the case in the formation of ester and peptide bonds which are of extreme biological importance. Catabolism and anabolism, if we limit these terms to imply the biological degradation and synthesis, respectively, of such a bond of 3000 calories, will differ fundamentally in the "detour" that must be resorted to by the living cell in order to create the energy-rich bond essential for anabolism. Thus the *breakdown* of the ester glucose-6-phosphate takes place by simple hydrolysis catalysed by the enzyme glucose-6-phosphatase. The *formation* of the ester from glucose and phosphate, however, requires an activation of the phosphate to the high-energy level in ATP, before the hexokinase reaction can take place. The thermodynamic relationships in such a process are shown diagrammatically in Fig. 26.

The mitochondria contain the enzyme complement of the cell for the conversion of the energy of the substrate into a form suited for the performance of work and for anabolic processes. In the detailed treatment of this enzyme complement in previous chapters, we have employed a classification corresponding to the different phases in the above diagram.

We have seen that the reactions in which respiratory energy is transferred to ATP and other energy-rich precursors essential for anabolic processes, together with the equilibrium reactions which constitute links between the precursors, are all mainly and characteristically located in the mitochondria. On the other hand, the exergonic reactions which, as an initial phase in the synthetic processes of the cell, utilize, or rather tap off, the energy of the precursors, cannot be said to be characteristically extra-mitochondrial or intramitochondrial. Evidently, this may quite simply be due to the fact that the sphere of activity of the mitochondria cannot be defined by thermodynamic considerations. The question remains, however, whether this apparent irregularity in the localization of the exergonic reactions provides adequate ground for the rejection of a thermodynamic boundary, which covers the facts so satisfactorily in the case of the first two types of reaction.

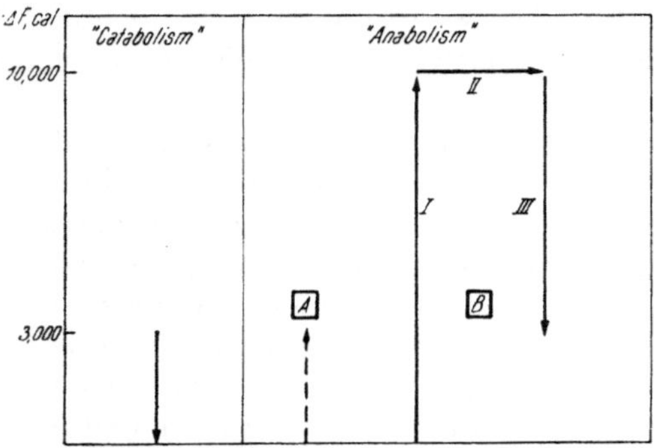

Fig. 26. Diagrammatic representation of energetics in catabolic and anabolic processes.

The terms "catabolism" and "anabolism" designate the hydrolysis and synthesis, respectively, of a 3000 cal.-bond (such as ester or peptide bonds). In the *catabolic* reaction, 3000 cal. are liberated. If the corresponding *anabolic* reaction were to proceed according to the same pattern, a supply of 3000 cal. would be needed to achieve synthesis (alternative A, dotted arrow).—In biological systems, however, the energy, derived from oxidative processes (alternative B, arrow I), is mediated by transfer through "high-energy bonds" of about 10,000 cal. (arrow II), the true synthesis taking place by an *exergonic* splitting of a high-energy bond (arrow III).

From purely theoretical considerations, we must assume that each exergonic enzyme reaction involves an enzyme-substrate complex at the high energy level. In this event, the first partial reaction, in which this complex is formed, is a transfer at the high energy level, while only the second reaction, in which the complex undergoes spontaneous decomposition, is actually exergonic. Enzyme-substrate intermediates are as a rule extremely unstable and hence the impression may be gained that the whole reaction is irreversible. Thus the fact that exergonic overall processes are found to occur in the mitochondria, does not necessarily imply that the enzymic activity of the mitochondria extends beyond a transfer at the high energy level.

2. Correlation between metabolic function and structure

If we now assume that the mitochondria are responsible for the "detour" which distinguishes anabolism from catabolism, we may enquire the reasons for a structural segregation of the anabolic function.

a) Coupling between oxidation and phosphorylation

May it be that the reactions in which respiratory energy is transferred to ATP and other energy-rich carriers—for the sake of simplicity, let us call them coupling reactions—cannot take place in an unorganized system? As it was observed some years ago that isolated particles contained not only the oxidative enzymes largely known at the time, but had the power, not possessed by a simple mixture of the enzymes, of coupling oxidation with phosphorylation, there was reason to suspect a relationship between structural organization and coupling power. This suspicion was reinforced by the finding that these particles not only brought about coupling reactions, but were also dependent for their oxidative power on the presence of the requirements for phosphorylation. It was also found that special measures had to be taken, often involving a disorganization of the structure, in order to uncouple respiration from phosphorylation.

More recently, largely as a consequence of the discovery of coenzyme A, the detailed mechanism of certain coupling reactions has been elucidated, and this has made it possible to bring about coupled phosphorylation with the aid of soluble enzymes. The recent work of GREEN, BEINERT, FULD, GOLDMAN, PAUL and SARKAR (1953) on oxidative phosphorylation in non-mitochondrial systems shows that it was not the absence of mitochondrial organization, but rather insufficient knowledge of the mechanism of the coupling reactions and lack of a suitable technique for the solubilization of the enzymes, which were responsible for earlier failures. We can agree with GREEN (1951 b) that it is only a matter of time until we shall be able to achieve complete coupling between oxidative reactions and phosphorylation in unorganized systems.

b) Efficiency of the organized mitochondrial system

Is, then, the purpose of an organized unit to make more efficient the processes taking place therein? This question arose as a direct consequence of GREEN's cyclophorase theory according to which, as has already been described, the organized cyclophorase system contains the full complement of coenzymes necessary in order to oxidize any substrate utilizable by the system to carbon dioxide and water, without accumulation of intermediates and at the maximum rate permitted by the enzyme complement. If the same oxidative capacity is to be attained in an unorganized system, the coenzymes must be introduced in quantities very much greater than those present in the organized system. In view of the fact that the quantities of these coenzymes present in living tissue are of the same order of magnitude as those found in cyclophorase preparations and certainly much smaller than are necessary to provide the same efficiency in an unorganized system, it appears to follow that the organized system is the most efficient attainable with the physiological quantities of enzyme and coenzyme. The validity of this reasoning is not affected by the manner in which the coenzymes are bound to the system, whether by conjugation or by membranes. This is a matter of dispute that has been discussed in an earlier section.

It would nevertheless be wrong to believe, as Potter, Recknagel and Hurlbert (1951) have already emphasized, that the cell, under physiological conditions, always works at the oxidative capacity made possible by the mitochondrial organization.

We shall illustrate this point with the aid of some examples.

Let us consider the system:

(1) α-ketoglutarate + DPN $\rightarrow$ succinate + DPNH$_2$ + CO$_2$
(2) α-ketoglutarate + DPNH$_2$ + NH$_3$ $\rightleftharpoons$ glutamate + DPN + H$_2$O

$$\overline{2\ \alpha\text{-ketoglutarate} + NH_3 \rightarrow \text{succinate} + \text{glutamate} + CO_2 + H_2O}$$

This sequence of reactions takes place if isolated mitochondria are incubated with α-ketoglutarate and ammonia under anaerobic conditions (Hunter and Hixon 1949 a). The process involves a dismutation over DPNH$_2$, *i. e.* one molecule of ketoglutarate is reductively coupled with ammonia at the expense of a second molecule of ketoglutarate, which is oxidized to succinate. The position of equilibrium in reaction (2) is determined by the initial concentration of ammonia. If, under otherwise identical conditions, we admit oxygen to the system, the DPNH$_2$ formed by the oxidation of ketoglutarate will be confronted by two alternatives for the surrender of its hydrogen. It can donate it for the formation of glutamate or to the respiratory chain for the formation of water. We can describe this as a competition for hydrogen between ketoglutarate and ammonia on the one hand and oxygen on the other. It has been shown by Still, Buell and Green (1950 a) that in the aerobic oxidation of succinate by isolated mitochondria, a concentration of ammonia even about one hundred times the physiological one is insufficient to compete with the oxidation of the substrate in favour of the synthesis of glutamate, and that the presence of a relatively high concentration of ketoglutarate *together* with the high ammonia concentration is essential for glutamate formation to compete effectively with respiration. This demonstrates that glutamate formation could not occur in the presence of physiological concentrations of ammonia and ketoglutarate, if the full oxidative capacity of the isolated mitochondria prevailed in the cell.

In a liver homogenate which oxidizes pyruvate in the absence of C$_4$-dicarboxylic acids, acetate is accumulated in the form of acetoacetate. If, however, a C$_4$-dicarboxylic acid is added as a source of condensing partner for the acetate, 90% of the pyruvate is oxidized to carbon dioxide and water, while 10% is still accumulated as acetoacetate (Potter and Heidelberger 1950). This implies that when there is free competition between the formations of citrate and acetoacetate, the former is favoured by a factor of about nine, and that all the factors which may inhibit the formation of citrate will favour the synthesis of acetoacetate. We may especially consider two such types of factors, those which restrict the supply of condensing partner and those which limit the oxidation and thus exert a secondary inhibition on the formation of citrate.

A ketogenesis can be brought about *in vitro* in two manners: a) by deficiency of condensing partner, achieved by omission of C$_4$-dicarboxylic acid or removal of this from the system by relatively high ammonia concentration or by malonate (RECKNAGEL and POTTER 1951), and b) by suppressing the oxidation of citrate through the introduction of anaerobic conditions, thus causing a secondary inhibition of citrate formation.

In pathological ketogenesis *in vivo* neither of these factors can be primarily responsible. Anaerobic conditions due to lack of oxygen can scarcely prevail. The ammonia concentration, even if its level may be raised, as in diabetics, can hardly be conceived as the primary agent in the ketosis. In what concerns the availability of condensing partner, there is reason to believe that the physiological carbon dioxide concentration is sufficient to ensure an adequate supply of oxalo-acetate by carboxylation of pyruvate (BARTLEY 1953). It therefore appears that pathological ketogenesis may be attributed to a factor which secondarily lowers the oxidative capacity. A further discussion of this question follows on p. 106.

To summarize these examples, we may say that a total combustion of substrates is antagonistic to the syntheses in which these are involved. The competition for hydrogen between oxygen and building blocks of synthesis is strongly in favour of the former at the full oxidative capacity of cell-free systems, *i. e.* homogenates or isolated mitochondria. In the intact cell. where a synthesis of glutamate *must* occur, and where a pathological keto-genesis demonstrably *can* occur, there must be a braking device whereby the high respiratory capacity may be limited.

It follows from this reasoning that the physiological purpose of the high oxidative capacity attainable in the mitochondrial organization, is not primarily to render the respiration more efficient. It serves rather to provide a range sufficiently wide to permit the functioning of those factors which determine in the living cell the ratio of substrate oxidized for the generation of energy to that preserved for synthetic processes.

We can hence assume that this regulatory mechanism or, to use the term employed above, braking device, is intimately related with the mitochondrial organization. We may expect that its attack will take place at the limiting factor in the organized respiratory system.

c) Reactions regulating the pathways of intermediary metabolism

Dinitrophenol, which uncouples the phosphorylation associated with the respiratory chain, increases respiration in intact cells (PEISS and FIELD 1948). This indicates that the factor limiting oxidation in the living cell lies in the reaction where phosphate is transferred to ATP from the primary ester formed in connection with the electron transport along the respiratory chain (cf. HUNTER 1951).

In isolated systems this limitation may be overcome in several manners. Mitochondria isolated in a salt medium have the power of accumulating inorganic pyrophosphate (CROSS, TAGGART, COVO and GREEN 1949) and also show a strong ATPase activity. These two properties, which may be considered to arise from a breakdown of the structure, make possible a rapid dephosphorylation of newly formed ATP and the maintenance of a relatively high concentration of ADP. which in turn contributes to a saturation of the limiting enzyme.

In mitochondria prepared in sucrose, the power of accumulating pyrophosphate and the ATPase activity are lowered or maybe absent (Kielley and Kielley 1951). If it is desired in this case to circumvent the bottle neck, ADP or adenylic acid must be added in sufficient concentration to ensure an adequate concentration of ATP. Apart from the fact that it is necessary to employ concentrations of adenine nucleotides greatly in excess of the physiological, which may be injurious to the system, this measure also involves the difficulty that the ATP accumulated in the system tends to equilibrate with ADP and adenylic acid via the myokinase reaction. Thus the situation arises that, whereas the limiting factor has been circumvented by the relatively high ADP concentration, the gradual diminution in the adenylic acid concentration restricts another reaction which was not previously limiting. This matter is more fully discussed on p. 64. A comparable situation, although less marked, occurs if we supplement a similar system with creatine and creatine transphosphorylase (Lardy and Wellman 1952), *i. e.* an enzyme which *reversibly* transfers the terminal phosphate group of ATP to an acceptor.

It follows that if a maximum phosphorylating capacity is to be obtained with a physiological concentration of adenine nucleotides, the system must be supplemented with an enzyme which ensures a continuous regeneration of ADP from ATP by an exergonic reaction, *i. e.* irreversibly. Such an enzyme is *e. g.* hexokinase which, in the presence of glucose, can actually bring the phosphorylating power of a mitochondrial preparation to a maximum (Kielley and Kielley 1951, Lardy and Wellmann 1952, Lindberg and Ernster 1952a).

It is difficult at present to decide whether, when the maximum rate of phosphorylation has been reached in such a system supplemented with hexokinase, the full respiratory capacity is actually being exploited. If isolated mitochondria are supplemented with isolated nuclei, the respiration increases just as when the hexokinase system is added. This has been supposed by Potter, Lyle and Schneider (1951) to be due to a tapping off of ATP by the ATPase of the nuclear fraction. Johnson and Ackermann (1953) have recently shown, however, that isolated nuclei have a stimulating effect upon the respiration, even when the system is simultaneously supplemented with hexokinase. The authors conclude that this additive stimulation is due to the presence in the nuclear fraction of an enzyme which phosphorylates ADP to ATP. If this is the case, it implies that the phosphorylating capacity of mitochondria is lower than the respiratory capacity even if the maximum removal of ATP takes place. The physiological significance of this effect of the nucleus is not yet clear. Brachet (1952a), who once presented good evidence for the direct participation of the nucleus in oxidative phosphorylation, has recently found (Brachet 1952b) that cell fragments free from nuclei retain their level of respiration several days after the enucleation and that the ATPase activity also remains unchanged.

The experiments with hexokinase-supplemented mitochondria have led to the recognition of the important fact that the respiration of the cell is regulated by the processes which utilize ATP, *i. e.* by the energy requirements of the cell (Potter, Recknagel and Hurlbert 1951, Lardy and Wellman 1952, Lindberg and Ernster 1952a). On the other hand, we have arrived above at the conclusion that respiration is antagonistic to those syntheses which require intermediates of the Krebs cycle as building units.

Hence the removal of ATP does not constitute a regulation of the respiration as such, but rather a regulation of the supply of Krebs cycle intermediates to the cell for synthetic purposes.

For a further study of this principle, we shall divide the substances partaking in anabolic reactions as defined earlier, into two groups: a) metabolites of the Krebs cycle and substrates which can be oxidized via the Krebs cycle without an exergonic intermediate reaction: to these we will apply the common term "mitochondrial substrates", and b) other

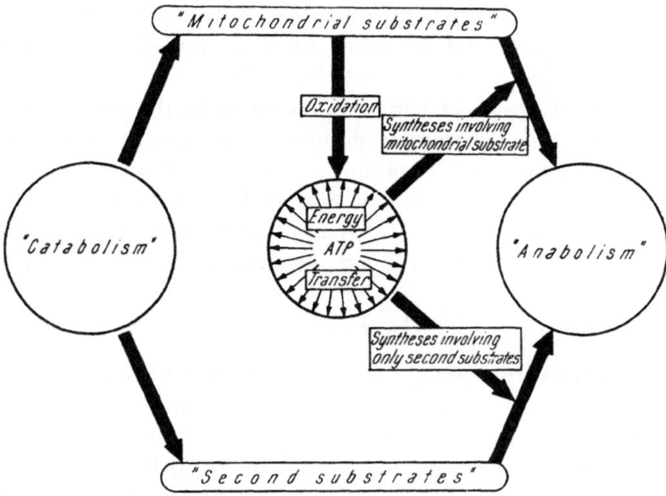

Fig. 27. Schematic illustration of the correlation between the availabilities of mitochondrial substrates as sources of energy and as building elements for synthetic processes.

The substances supplied to the cell by catabolic processes are divided in the figure into a) substrates oxidizable without any intervention of exergonic reactions ("mitochondrial substrates"), and b) other metabolites which may serve as building elements in synthetic processes ("second substrates"). In a mitochondrial system, coupled with a synthetic reaction involving only "second substrate", any "mitochondrial substrate" supplied to the system will be wholly available as source of energy for the synthesis, and the extent of its combustion will be determined by the energy requirement of the synthetic process. When, as a second alternative, a synthetic reaction involving some "mitochondrial substrates" is coupled to the mitochondrial system, the extent of oxidation will be "balanced" with that of synthesis in such a manner as to yield equivalent quantities of building elements and energy. In the living cell, where both types of synthetic processes are present, the extent of combustion of "mitochondrial substrates" will, therefore, be determined by the extent of the synthetic processes only involving "second substrates".

metabolites, which, for the sake of simplicity, we will call "second substrates". We can accordingly classify anabolic reactions in which oxidative energy is utilized in two groups: a) reactions in which "mitochondrial substrates" are involved as building units, and b) reactions in which only "second substrates" participate.

If we commence with a certain quantity of a "mitochondrial substrate" S and have a single anabolic reaction of type b, i.e. a reaction involving an exergonic coupling of two "second substrates", it is evident that the substrate S will all be available for the provision of oxidative energy for this exergonic reaction. Let us now consider the case that we have an anabolic reaction of type a, where coupling takes place either of a "mitochondrial substrate" (S itself or a derivative thereof) with a "second substrate", or of two "mitochondrial substrates" with one another. In this case it is clear that the oxidation of S must be limited in such a manner that the substrate or substrates required for the anabolic process are preserved from oxidation. Since, on the other hand, the oxidation is limited, as has been shown above, by the tapping off of energy—in this case by the anabolic reaction in question—the character of this reaction will determine the quantity of S that is oxidized.

In a physiological system it is clear that both these types of anabolic process will occur. It follows from the above reasoning that the reactions of type b, $i.e.$ those in which "mitochondrial substrates" are not involved, will determine whether, and to what extent, "mitochondrial substrates" are made available for syntheses (see Fig. 27).

These anabolic reactions of type b correspond to the group of exergonic reactions that we have formerly described as kinase reactions in the limited sense (reactions involving an exergonic splitting of ATP). The braking mechanism for respiration, which we believe to be necessary if the cell is to utilize "mitochondrial substrates" for synthetic purposes, must therefore be sought among the factors regulating the kinase reactions.

d) Hormonal regulation of kinase reactions

A kinase reaction, the physiological regulation of which has been intensively studied in recent years, is the glucokinase reaction. In 1939, Lundsgaard found that the uptake of glucose by perfused muscle preparations could be considerably stimulated by insulin. He suggested that the glucose could not enter the cell by simple diffusion and that the site of action of the insulin was the cell surface. Cori and his co-workers (Cori 1949), in experiments with isolated rat diaphragm, have demonstrated that the stimulating effect of insulin on the uptake of glucose may be attributed to an activation of the glucokinase reaction. It is now recognized (cf. Colowick, Cori and Slein 1947, Walaas, Walaas and Løken 1952, Bullough 1952, 1953) that the hormonal control of the glucokinase reaction is not effected only by insulin but also by a number of other hormones. The oestrogenic hormone, for example, acts in the same sense as insulin, whereas corticosterone and pituitary growth hormone inhibit the glucokinase reaction.

The physiological significance of this control mechanism has been elegantly demonstrated in the experiments of Bullough (1952, 1953) on the factors determining mitotis activity in mouse epidermis. Intact cells of this tissue lack endogenous substrate and cannot undergo mitosis in a buffered salt solution unless oxidizable substrate is added. If glucose is employed as substrate, a marked hormonal effect upon the mitotic activity can be demonstrated: insulin and oestrogenic hormone stimulate the activity, whereas the pituitary growth hormone and corticosterone have an inhibiting action. If substrates other than glucose are used, such as fructose or lactate, the maximal mitotic activity is obtained irrespective of the supply of hormones. From these observations, BULLOUGH concludes that the glucokinase reaction is the site of action of the hormones and that the mitotic activity is determined by the degree in which the hormonally controlled glucokinase permits the cell to take up glucose for the generation of energy for the mitotic process.

The dependence of mitotic activity upon glucokinase activity and the dependence of the latter upon the hormonal balance have been placed beyond doubt by the experiments of BULLOUGH. They also provide an ample confirmation of the earlier suggestions (see above), according to which glucokinase regulates the uptake of glucose by the cell. We are not justified, however, in concluding from these experiments that the hormonal control of the glucokinase reaction is *solely* concerned with the regulation of the glucose uptake.

It appears to be fairly well established that structural integrity is an absolute requirement for hormonal effects on glucokinase (cf. CORI 1949). It has therefore been assumed that the hormonal control only affects surface-bound glucokinase. Suggestions have even been made that hormones such as insulin cannot enter the cell (ROSENBERG and WILBRANDT 1952). In other words, structural integrity of the cell has been quite generally taken to imply an intact surface. The requirements of structural organization may, in fact, be fulfilled when the cell surface is no longer intact.

This is illustrated by the experiments of CORI and his co-workers (cf. CORI 1949) on the action of the hyperglycaemic-glycogenolytic hormone of the pancreas (H-G factor) on the phosphorylase reaction. This hormone stimulates the limiting reaction in the reaction chain glycogen → glucose. where glycogen is split phosphorolytically to give glucose-1-phosphate. If liver slices are incubated in the presence of the H-G factor, an increased breakdown of glycogen is obtained as compared with a control experiment. This effect of the H-G factor cannot be obtained if the cell structure is destroyed by, for example, freezing and thawing. It is clear in this case that the structural essentials are not the cell surface and permeability factors.

CRANE and SOLS (1953), as already described, have found that the animal glucokinase is largely bound to a particulate cytoplasmic fraction with a sedimentation rate identical with that of the mitochondria. The enzyme differs from the soluble hexokinase of yeast in that it is inhibited by extremely low concentrations of its reaction product. glucose-6-phosphate. The authors consider it highly likely that such an enzyme could be governed hormonally by a regulation of the prevailing concentration of glucose-6-phosphate.

We have come to the conclusion that a regulation of the activity of the kinase reactions might form the basis for a regulation of alternative pathways in mitochondria. It therefore does not appear to be excluded that a hormonal regulation of glucokinase reactions might also occur at the mitochondrial level of organization.

Furthermore, if we accept the general thesis that processes utilizing ATP exert control over the balance between synthesis and respiration, we might expect hormonal regulation to be present not only for the glucokinase reaction, but also for other reactions where ATP is irreversibly split. Some evidence for this is found in the report of Lardy (1952), according to which thyroxine activates the breakdown of ATP in isolated mitochondria. The mechanism of this effect has not yet been elucidated.

e) Summary

The above considerations of the physiological significance of the mitochondrial organization may be briefly summarized as follows:

1. The mitochondria contain the enzyme complement of the cell for the conversion of the energy of the substrate into forms suitable for anabolic processes.

2. The oxidation of the different substrates involves a generation of energy, but simultaneously a degradation of building elements essential for certain syntheses.

3. The mitochondrial structure enables the cell to display the maximum oxidative capacity attainable with its complement of enzymes and coenzymes.

4. This maximum oxidative capacity makes possible an extensive control over the essential balance between the oxidative generation of energy and the availability of material for synthesis.

5. The limiting factor in cell respiration is the phosphorylation associated with this, the intensity of which is determined by the activity of the reactions utilizing ATP. It is known that certain reactions of this type are subject to hormonal control in the intact structure.

3. Specialization phenomena in mitochondria

In the preceding section we have discussed the general principles which are believed to govern the enzymic organization of the mitochondria. We have attempted to identify the factors in this system which serve to regulate the basic activity of the mitochondria, adapting it to the requirements of the individual cell and tissue. On this basis we have gained some idea of how, in a cell characterized by a vigorous synthetic activity, the "technical" problem of preventing the total oxidation of the substrate may be solved, or how fluctuations in the generation of energy within a cell can be adapted to the current energy requirements.

In the formulation of these fundamental principles of metabolic control, we have not found it necessary to postulate a differentiation among mitochondria. The mere fact, however, that the mitochondria are able to convert oxidative energy into *special* forms for divers anabolic processes, suggests that they do not only obey the orders from without, which determine the degree of oxidation, but harmonize with the metabolic character of the cell through their own enzymic processes, the activation reactions. All mitochondrial reactions, in fact, with the exception of those involved in the Krebs cycle and in the formation of ATP and certain amino acids, appear to be more or less characteristic features of special varieties of mitochondria. In this chapter we propose to make a survey of the properties of specialized mitochondria.

a) Behaviour of mitochondria during embryonic development

An illuminating example of mitochondrial differentiation and its relation with cell differentiation is provided in the studies of Gustafson and Lenicque (1952) and Gustafson (1952) on the rôle of the mitochondria during the embryonic development of the sea urchin. In the immature egg of the sea urchin there is vigorous mitochondrial activity which is believed, on the ground of morphological observations, to be associated with the metabolic events leading to the ripening of the egg (Runnström, personal communication). In the mature egg very few, if any, of the mitochondria of the oocyte are to be found, while several enzymes, such as cytochrome oxidase, which are otherwise typical particulate enzymes, are found here to be dissolved in the ground cytoplasm (Hutchens, Kopac and Krahl 1942). In these eggs it is possible to demonstrate respiration in the surface layer (Warburg 1910), and Runnström (1952) believes that this membrane is largely responsible for the respiration of the egg and its energy supply. This common characteristic of the cell membrane and the mitochondria is not, indeed, confined to the sea urchin egg, and it has led Runnström to regard the cell surface as an organite in some degree comparable with the mitochondria.

The fertilization of the sea urchin egg is followed by a pronounced increase in respiration (Warburg 1908, 1910), although there is no sign of a corresponding formation of mitochondria. It would thus seem that the conclusion of the period of determination, which occurs during the blastula stage with the appearance of the first mesenchyme cells. a new stage with the appearance of the first mesenchyme cells, a new increase in respiration commences (Lindahl 1939) which is a direct function of a mitochondriogenesis (Gustafson and Lenicque 1952). At the commencement of the invagination of the entoderm, the increase in respiration ceases. together with the rise in the mitochondrial count; at this stage there is a definite gradient in the mitochondrial density along the animal-vegetal axis of the larva (Fig. 3, p. 12).

During the period of differentiation, according to Gustafson and co-workers (Gustafson and Hasselberg 1951, Gustafson and Lenicque 1952, Gustafson 1952), the ectodermal mitochondria have an especially prominent rôle. All events during gastrulation in the ectoderm are a function of the mitochondrial activity (Fig. 28). The mitochondria appear to have mainly synthetic functions. These are manifested in the development of a secretion which forms the apical tuft of the larva and in the synthesis of proteins necessary for cell stretching and for the genesis of new mitochondrial populations.

The mitochondria of the entoderm, which apparently assume a more passive rôle during this period. are nevertheless engaged in syntheses.

especially of pigments and sulphate esters. The latter are presumably formed in detoxication processes (Lindahl 1936). Furthermore, a substance is formed at the vegetal pole, which inhibits the animal development.

The vegetal metabolism is characterized by a sensitivity to sulphate deficiency and to SH-inhibitors. Lack of sulphate results in a cessation of detoxication processes, cytolysis in the vegetal half, temporary hyper-animalization and ultimate death. SH-reagents seem to inhibit the specific inhibitor which serves to suppress the animal development. They hence produce a strong animalization of the larva.

The animal metabolism can be specifically inhibited with Li^+. The mechanism of this effect is unknown. Lindahl and Kiessling (1952) have

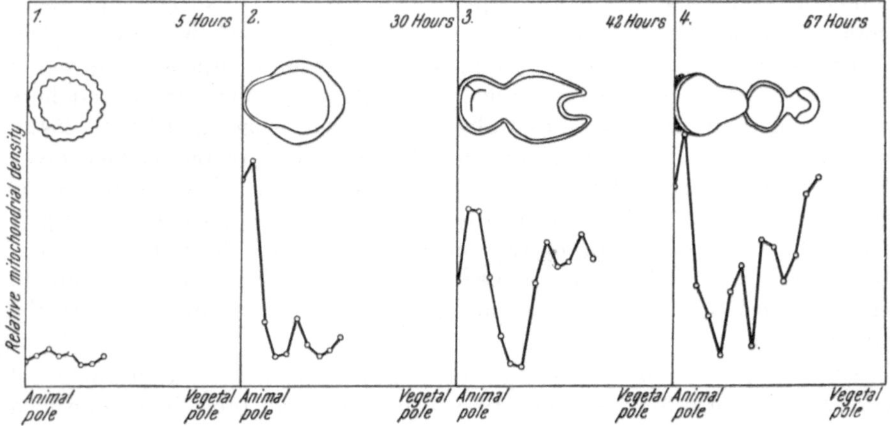

Fig. 28. Correlation between relative mitochondrial density at various levels of the animal-vegetal axis and different stages of differentiation in the larva of the sea urchin, *Psammechinus miliaris*.
(From Gustafson and Lenicque 1952.)
1) Blastula; 2) incipient gastrulation; 3) and 4) later stages of gastrulation. The larvae have been treated with Li^+ which causes over-vegetalization resulting in a protrusion, rather than invagination, of the entoderm. Li^+-treatment thus provides a good possibility of examining the mitochondrial density in different regions of the larva.

shown, however, that larvae poisoned with lithium ions accumulate inorganic pyrophosphate. Since an accumulation of pyrophosphate may be considered to be symptomatic of injury to the mitochondrial structure, this finding is of interest in that it suggests a possibility of correlating specific synthetic character with mitochondrial structure.

b) Muscle mitochondria

Further information regarding the differentiation of mitochondria during embryonic development has recently been gained from the studies of Lenicque (1953) on the development of somatic muscle in amphibia. In the early muscle tissue there is a fairly evenly distributed and homogeneous population of small mitochondria ("petites mitochondries"). In the course of the development, it is observed that this population undergoes differentiation in two senses. One type of differentiation leads to the formation of rod-shaped elements which, according to Lenicque, play a part in the lipid metabolism. Other small mitochondria are arranged along the length of the fibrils and serve as sources of ATP for the contraction process.

This latter variety of mitochondria, which are usually called sarcosomes,

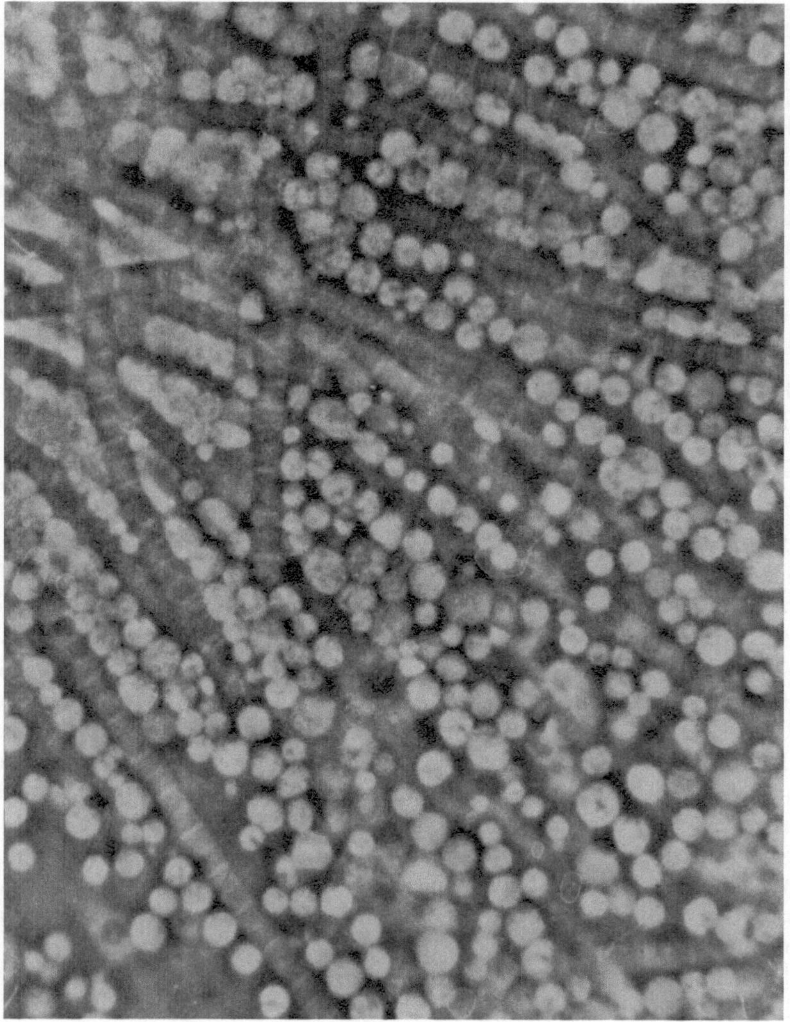

Fig. 29. Muscle fibrils with adjacent rows of sarcosomes. Flight muscle of *Drosophila funebris.*
Phase contrast. Magn. 2000 ×.
(From WATANABE and WILLIAMS 1951.)

have been studied in detail by WATANABE and WILLIAMS (1951). The sarcosomes in the flight muscles of *Diptera* are giant structures (up to $4\,\mu$ in diameter) arranged in rows along the fibril and enclosed in the sarcolemma (Fig. 29). They constitute not less than one third of the muscle mass and may be very readily isolated by differential centrifugation. When isolated, they show a high respiration in the presence of intermediates of the Krebs cycle. Sacktor (1953), who has worked with isolated sarcosomes of the housefly, has nevertheless recently reported that he was unable to obtain coupled phosphorylation with this material, even in the presence of yeast hexokinase and glucose as phosphate acceptor system.

It is quite clear from the morphological evidence that the sarcosomes are specialized for the formation of ATP essential for muscular contraction. It is still difficult to understand how in this case the respiration can be immediately controlled by the ATP-utilizing reactions, which are apparently limited here to the splitting of ATP by the actomyosin complex. It might be suggested that the creatine transphosphorylase system acts here as a controlling buffer mechanism, since a relatively high creatine level can ensure a continuity of respiration even during rest.

c) Liver mitochondria and processes concerned with metabolism of fatty acids and with detoxications

The studies of Gustafson provide a clear indication that the mitochondria can be differentiated for various metabolic functions. Not only, apparently, are they capable of being controlled by external factors as regards the extent to which the oxidative energy is used for the formation of ATP or for the activation of substrates: they also have the power of specifically diverting energy, within a given range, for certain processes characteristic of the cell. We have several concrete examples of this from biochemical studies of the mitochondria of different tissues, even though it is not possible as yet to fit them into the general physiological picture.

We find, for instance, that there are enzymic differences between mitochondria of different organs with regard to the oxidation of fatty acids (cf. Green 1952). Liver mitochondria lack the enzyme system necessary for the oxidation of acetoacetate. Other organs, with the exception of brain, possess in their mitochondria the complete apparatus for the total oxidation of the physiologically occurring fatty acids, *i. e.* those with even numbers of carbon atoms. Brain mitochondria apparently lack all power of attacking fatty acids (Brody and Bain 1952). It is a matter for future investigation to determine whether these facts may lead to an understanding of the physiological coördination of fatty acid metabolism in multicellular organisms and of the key position of the liver in this metabolism.

The powerful hierarchy which prevails in multicellular organisms in the syntheses of nitrogenous end products and in other detoxication processes, appears to be founded on differentiation at the mitochondrial level. The synthesis of urea in mammals, which takes place by way of the ornithine cycle, previously described in detail, is confined specifically in its first stage to the liver mitochondria (Müller and Leuthardt 1949). The conversion of the citrulline thus formed to arginine can occur in both liver and kidney (Leuthardt 1952). The final stage, the arginase reaction, is ubiquitous (Schein and Young 1952).

Detoxications of aromatic substances are reactions typical of the liver mitochondria. As we have described in an earlier chapter, the mechanism of these processes consists in the coupling of the aromatic molecule to a radical—sulphate, acetate, glycine or glucuronate—brought into an active form.

In considering the many types of specialization of which mitochondria are capable, we are confronted by the problem of whether all these properties may be displayed by one and the same type of mitochondrion, or whether there are several types of particles, each with its own speciality. It is fairly evident in the case of muscle that different varieties of mitochondria are present in the same tissue, and the morphological data show clearly that the differences are determined by adaptation to different functions. We have no definite information as to the state of affairs in liver. It nevertheless seems reasonable to suppose that the synthesis of citrulline and the detoxication of phenol, for instance, are not entrusted to the same type of mitochondrion. It would seem rather senseless, physiologically speaking, that these two processes should compete for the energy from the same energy-generating unit.

Reasoning along these lines may, however, lead to an unwarranted degree of schematization, especially in view of our present knowledge of the mechanism of tubular reabsorption in the kidney.

d) Rôle of mitochondria in renal reabsorption

It has been known for some time that competition occurs in the renal reabsorption between different sugars, different amino acids, etc. It was also suggested at an early stage that this effect consisted in a competition for a common carrier in the kidney cells specialized for the accumulation of the reabsorbed substances (SHANNON and FISHER 1938, BEYER, WRIGHT, SKEGGS, RUSSO and SHANER 1947). A series of recent investigations by TAGGART and his co-workers have thrown some light on this mechanism. Experiments were made with isolated mesonephros of the flounder, where the conditions for the accumulation of phenol red could be studied (TAGGART and FORSTER 1950, FORSTER and TAGGART 1950). It was shown that agents such as dinitrophenol, which uncouple respiration from phosphorylation, inhibit the accumulation of phenol red in the cells. This finding indicates that oxidative phosphorylation is essential to the accumulation mechanism; in other words, the process is intimately associated with the mitochondria.

In another series of experiments, TAGGART and co-workers (MUDGE and TAGGART 1950 a, b, CROSS and TAGGART 1950) have studied the accumulation of p-aminohippuric acid in slices of dog kidney. Here also it was shown that oxidative phosphorylation was necessary. Further information was obtained regarding the mechanism for the active formation of p-aminohippuric acid. The capacity of accumulation is raised if acetate is added to the medium and considerably reduced if the acetate is replaced with a dicarboxylic acid. This indicates that the condensing reaction in the

Krebs cycle is antagonistic to the reaction in which p-aminobenzoic acid is coupled to glycine. This may be a further example of the phenomenon which we have previously encountered in the case of liver, where keto-genesis and oxidation are competitive.

It has been shown by ZOLLINGER (1950) on the morphological level that the kidney mitochondria constitute a uniform population with a wide capacity for different accumulative processes. If ovalbumin, haemoglobin (as blood haemolysate). gelatin or sucrose is injected intravenously or intraperitoneally into an animal, the test substance is accumulated in the convoluted kidney tubules, where it is taken up by the mitochondria. This process can be readily followed under the phase-contrast microscope. The mitochondria accumulate the substance with a large increase in volume and effect its removal from the cell in the form of secretion droplets.

ZOLLINGER has followed the accumulation process in great detail. Ovalbumin, haemoglobin and vital stains, the latter accompanied by plasma proteins, first appear as small scattered droplets within the mitochondrial core. These sub-sequently coalesce and undergo a distinct chemical change. It was shown, for example. that the accumulated ovalbumin could be stained between 24 and 40 hours after the injection, but not earlier or later, with specific fibrin stains. The accumulated material finally occupies the entire mitochondrial body, where-upon the mitochondrial membrane degenerates and the mitochondrion becomes a drop of secretion. In the case of sucrose the process follows a somewhat different course. The accumulation commences in the space between the mem-brane and the mitochondrial core and penetrates inward, finally, as in the previous case. displacing the mitochondrial structure. No chemical change was observed in the case of sucrose.

e) Specialization of mitochondria for cellular accumulation and secretion

The above observations led ZOLLINGER to attribute an accumulative function to the mitochondria. This was, indeed, one of the earliest sug-gestions regarding the physiological rôle of the particles (REGAUD 1909). Good evidence in favour of the hypothesis was obtained when GUILLIER-MOND found in 1932 that the plastids of plants derive from mitochondria. ROJAS and DE ROBERTIS described in 1936 how the mitochondrial content of maturing amphibian erythrocytes decreased in parallel with the syn-thesis of haemoglobin. The fully developed erythrocyte contains no mito-chondria. DE ROBERTIS, NOWINSKY and SAEZ (1948) made the analogous observation of an inverse relationship between the number of mitochondria on the one hand and the numbers of fat droplets and glycogen granules on the other in amphibian liver in different physiological states.

With the aid of the method described above, ZOLLINGER has shown that mitochondria in organs other than kidney can also be made, under suitable conditions. to accumulate different test substances. He has also been able to demonstrate that the events in a secreting cell constitute a physio-logical variant of his own model experiments. In mast cells and the ex-cretory cells of the pancreas. he has observed distinctly how the mitochondria, which at the beginning of their secretory activity are oriented basally in the cells, engage in the formation of the specific cell secretions, in this case heparin and zymogen respectively. The mitochondria swell during the accumulation of these products, and are finally converted into droplets of secretion. During this process the particles migrate to the apex of the cell where they are finally excreted by an unknown mechanism.

ZOLLINGER believes that this represents the general mechanism for the formation of cellular secretions and inclusions. It provides a reasonable picture of the origin and nature of secreted morphological structures such as the previously mentioned apical tuft of the sea urchin larva, and of granular cell inclusions such as pigmented granules, fat droplets, cortical granules etc.

f) Summary

By way of summarizing the phenomena connected with mitochondrial specialization, the following tentative scheme may be put forward:

Each mitochondrion is furnished with a basic complement of enzymes which catalyze the reactions of the Krebs cycle, the activation of the intermediates of that cycle, the coupling between oxidation and phosphorylation and the formation of ATP.

The reactions associated with the metabolism of fatty acids, the synthesis of nitrogenous end products, the condensation for purposes of detoxication of aromatic substances with activated radicals, and the active formation of substances to be accumulated or secreted, are special functions based on the fundamental pattern.

There appears to be a principle of competition between processes occurring in one and the same type of mitochondrion. The basis of this competition may consist in the relative activities (perhaps hormonally regulated) of the different enzymes and the availability of the necessary substrates. The mitochondrion will thus represent a versatile and continuously working factory.

The complete specialization encountered in secretory mitochondria may be explained by the formation of a product which is accumulated within the particles. Such specialized mitochondria have been led into an activity which can no longer proceed on the basis of continuity, but which must needs be progressive in view of the indiffusibility of the product. This is best illustrated by the case of kidney mitochondria in which the accumulative activity can proceed continuously and over a wide range, involving a large variety of substances, as long as the particles are not encumbered with a substance which cannot leave the mitochondrion in the accumulated form but fills it and is only ejected after its destruction.

4. Relation of some pathological phenomena to mitochondrial function

Some of the above considerations may possibly lead to an understanding of the biochemical background of certain pathological states, by attributing the latter to disturbances in mitochondrial processes. Cases may be found where there is a disturbance in the control mechanism for alternative pathways, while in other instances the histopathological symptoms are due to a degeneration of the mitochondria in a tissue.

a) Aberrations in the regulation of alternative metabolic pathways

The biochemical aberrations in diabetes may be referred to displacements in the balance of hormones controlling the glucokinase reaction. A reduction in the rate of this reaction will result in a reduced uptake of glucose from the blood by the different tissues, with a consequent elevation of the blood sugar level. There will also be a reduction in the oxidative capacity which, in liver mitochondria, will give rise to an increased synthesis of acetoacetate. It should be pointed out that it would be difficult to accept an inadequate uptake of glucose as the sole consequence of a reduced glucokinase activity and thus as the only basis of the diabetic syndrome. A more promising line of thought is offered by the suggestion that the glucokinase reaction also acts as a regulatory mechanism in the control of the "oxidation-synthesis switch" in the mitochondria. It may be possible in this manner to explain that shifting of alternative pathways, which in the present instance favours the formation of acetoacetate.

Another much discussed disturbance in the balance of alternative pathways is found in the metabolism of neoplasms. POTTER and HEIDELBERGER (1950) believe that neoplastic growth can be attributed to two factors: a relative deficiency of oxidative enzymes, resulting in an increased supply of the building units essential for protein synthesis, together with a hyperactivity of enzymes involved in the synthesis of proteins and nucleic acids.

It is generally accepted that cancer cells have a lower oxidative capacity than normal cells, in addition to a very high glycolytic activity. Carbon compounds are thus made available for synthesis in large quantities (cf. ZAMECNIK 1952). The low activity of the oxidative system was formerly thought to be due to a deficiency in condensing enzyme (POTTER and BUSCH 1950), but it has recently been necessary to abandon this view (WENNER, SPIRTES and WEINHOUSE 1952). In fact, there appears to be no qualitative difference in the complements of oxidative enzymes in normal and tumour cells (WEINHOUSE 1951). Experiments with tumour homogenates have nevertheless shown inadequacies in the complement of coenzymes, due either to lack of these substances or to inefficient organization (WEINHOUSE 1951, WEINHOUSE, MILLINGTON and WENNER 1951, POTTER and LYLE 1951, WENNER and WEINHOUSE 1953). However this may be, it is considered likely that the oxidative capacity prevailing in cancer tissue will not be able to compete with synthetic processes involving reduction, such as reductive amination.

Reasoning along these lines helps us to understand how an increased synthetic activity may be effected by a tissue. It does not, however, ex-

plain why this synthetic activity should, in neoplastic tissue, be especially
concerned with protein and nucleic acid. The hyperactivity of the
enzymes catalysing these syntheses hardly suffices as an explanation, since
tumours in animals fed on a diet deficient in nitrogen still grow at the
expense of other tissues (cf. review by MIDER 1952). It appears that
growing tissues have an especial tendency to form precursors of proteins
and nucleic acids (cf. ZAMECNIK 1952) by mechanisms which are independent
of the degree of activity of the enzymes responsible for the continuation
of the syntheses. These precursors, at least in the case of protein synthesis,
may be mitochondrial products (SIEKEVITZ 1952).

It does not appear unreasonable, therefore, to suppose that the mitochondria contribute in two respects to the biochemical aberrations of cancer
tissue: a) they show a deficient oxidative capacity and can hence contribute large amounts of substrates and hydrogen for synthetic purposes;
b) they are specialized for the formation of precursors suitable for the
synthesis of proteins.

b) Mitochondrial degeneration

The investigations by ZOLLINGER (1950) of the accumulative power of
mitochondria have made it possible to interpret the histological picture
which characterizes the syndrome associated with parenchymatous degeneration. The "cloudy swelling" observed in nephrotic kidney tissue
is essentially the same phenomenon as that induced in ZOLLINGER's experiments by intravenous injection of ovalbumin. Similarly, the fatty
degeneration of the liver which is a fatal consequence of the action of
various toxic substances (phosphorus, diphtheria toxin, etc.), is a pathological accumulation of fat droplets in mitochondria. Whether in this
latter case the process is a chemical transformation of substances rich in
fat or merely an accumulation of lipids, is still obscure. At all events,
it seems clear that parenchymatous degeneration is due to a pathological
specialization of mitochondria in the accumulation of certain materials,
which renders impossible the continuation of their normal function.

B. Microsomes

In the foregoing attempt to form a picture of the physiological function
of the mitochondria, we had the considerable advantage of being able to
consider, from the outset, the isolated mitochondrion as a separate unit.
This unit could be studied not only in the morphological and chemical
respects, but was also suited to enzymological studies as a quite independent system, readily separable from other functional systems of the cell.
When a basic scheme had thus been drawn up, it was possible to consider
the reasons for the structural segregation of the mitochondrial system and
the manner in which it is integrated with its cellular environment.

It appears difficult to apply a similar procedure in studying the physiological function of the microsomes. As we have seen when discussing
their chemical constitution, the available data concerning the microsomes

are all relative in nature. It might be possible to express their dimensions in absolute terms, but even here there has been considerable divergence of views during the past years. Several sceptical pronouncements have been made, ranging from suggestions that the microsomes do not represent a clearly defined cell constituent (Chantrenne 1947, Barnum and Huseby 1948, Jeener 1948, 1952) to direct claims that they are, in fact, degradation products of mitochondria (Green 1951 a, b). This latter view may, however, be confidently written off. It may be argued, for example, that the high content of RNA in the microsomes excludes their derivation from the mitochondria, which are poor in RNA. On the other hand, a heterogeneity among the microsomes is insufficient ground for scepticism regarding their fundamentally uniform physiological function. The same scepticism would, indeed, have been justified with regard to the mitochondria, in which the differentiation phenomena might superficially have appeared to represent heterogeneity.

The firmest ground, relatively speaking, for the argument that the microsomes have a clearly defined function, lies in their characteristically high contents of RNA and of lipids. The data are, let it be repeated, relative, indicating that about 50% of the RNA and 50% of the lipids of the cell occur in the microsomes, constituting about 12% and 40% respectively of their weight. The remainder consists principally of proteins, among which some enzymes of phosphatase character have been identified, although it has not been possible as yet to determine the physiological significance of these. It is also difficult to assess the functions of the lipid components which, in view of the inadequacy of the chemical methods at our disposal and the poverty of our knowledge regarding the rôle of lipids in metabolic processes, cannot be distinguished from the lipids of the remainder of the cell.

The prevailing view of the function of the microsomes in the cell is thus based exclusively on the fact that the greater part of the cytoplasmic nucleoproteins are bound to these particles. It has accordingly been thought justified to suppose that the microsomes have functions in protein synthesis and in the maintenance of the genetic continuity of the cytoplasm.

1. Rôle of microsomes in protein metabolism

The belief that the microsomes partake in protein synthesis arises from the fact that the RNA content of the cytoplasm in a cell is correlated with the intensity of protein synthesis (Brachet 1940, 1941, Caspersson 1941, 1947). The best observations have been made on cells engaged in intensive protein synthesis, such as strongly proliferating bacteria, regenerating and neoplastic tissues and cells producing protein-containing secretions. The correlation between RNA content and protein-synthesizing activity holds without exception in all cases investigated, provided that the system is not deficient in some respect, e. g. in the supply of phosphate or nitrogen.

The proof of this correlation, which was initially quite empirical, has been facilitated by the circumstance that the protein, in the synthesis of which the microsomes are believed to participate, appears to be retained bound to the microsomes during this process. If, therefore, labelled amino acids are administered to an animal, the microsomes show the highest specific activity in the protein of all the cell fractions (BORSOOK, DEASY,

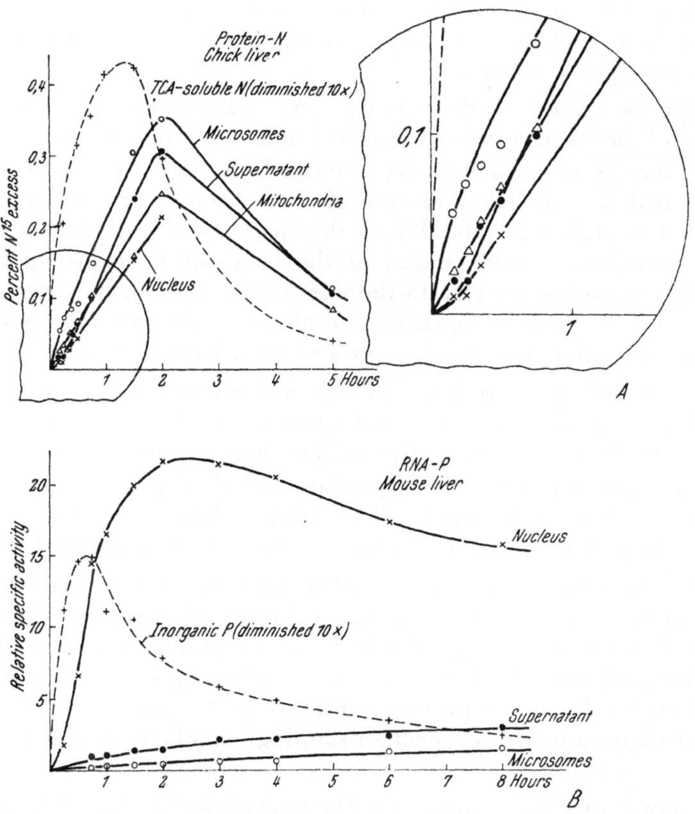

Fig. 30. Turnover *in vivo* of protein nitrogen and RNA phosphorus in liver cell fractions. *Upper figure:* Incorporation of N¹⁴-glycine into liver fractions of newly hatched chicks. (After HULTIN 1950.)
Lower figure: Incorporation of Na₂HP³²O₄ into RNA separated from mouse liver fractions. (After BARNUM and HUSEBY 1950.)

HAAGEN-SMIT, KEIGHLEY and LOWY 1949, HULTIN 1950, KELLER 1951, SIEKEVITZ and ZAMECNIK 1951). (See Fig. 30 A.)

The empirical relationship between microsomes and protein synthesis was brought on to a plane more comprehensible to the biochemist when BORSOOK, DEASY, HAAGEN-SMIT, KEIGHLEI and LOWY (1950) showed that isolated mitochondria and microsomes incorporated *together* more C¹⁴-glycine in their protein than did either particulate fraction alone. SIEKEVITZ (1952) has subsequently made a more thorough study of this system. He finds that isolated microsomes can incorporate C¹⁴-alanine in their protein, provided that they are supplemented with respiring mitochondria or with a low-molecular factor, stable against hydrolysis in normal acid for 7 minutes, which is obtained when isolated mitochondria are incubated with α-ketoglutarate. Addition of hexokinase and glucose to the mitochondrial system inhibits the formation of this factor. The experiments of Siekevitz, providing as they do the most concrete evidence yet available of the participation of microsomes in protein synthesis, show also that the mitochondria are involved in this process by their contribution of a low-molecular precursor.

Although we do not wish to enter upon wild speculation concerning the nature of this precursor, we can exclude with reasonable safety the possibility that it is a high-energy intermediate, in view of its stability against hydrolysis. It may, as has already been suggested (p. 83), be a γ-glutamyl peptide. Should this be the case, the rôle of the microsomes in protein synthesis would consist in the catalysis of transpeptidations. If we confine ourselves strictly to the experimental data, however, we can merely say, on the basis of Siekevitz's results, that the microsomes catalyse an exchange of amino acid residues in a protein bound to them.

The nature of the catalysts in the microsomes has not yet been elucidated. A quite sensational contribution in this field has recently been made by Binkley (1952). This author has isolated a preparation of nucleic acid, apparently free from protein, and shown that it catalyses the hydrolysis of cysteinylglycine. The preparation is specific for the *l*-peptide and is dialysable, indicating a relatively low molecular weight.

Even if this finding requires further confirmation, it indicates that nucleic acid is involved in the enzymic principle of the microsomes. The fundamental phenomenon that the intensity of protein synthesis in a cell can be correlated with the RNA content of the cytoplasm, will thus imply that the limiting factor in protein synthesis is the supply of ribonucleoproteins in their capacity of "protein-shaping" catalysts within the microsomes.

Let us now return for a moment to the part of the mitochondria in protein synthesis. We have supposed above, with a good basis of fact, that this part consists in the production of certain "key peptides" which act as precursors for the transpeptidating and protein-shaping activity of the microsomes. Gustafson, Hjelte and Hasselberg (1952) have succeeded in demonstrating a progressive appearance of peptides of increasing length during the early development of sea urchin larvae. The production of these peptides can be retarded by treatment of the larvae with lithium ions. Since the site of attack of the lithium is probably the mitochondria (cf. p. 99), it appears that these peptides derive from mitochondrial activity. The isolated peptides, as Gustafson and his co-workers have shown, constitute a growth factor for *Lactobacillus casei*. Hence the formation of the peptides might well be a limiting factor in the protein synthesis. The recent studies by Snellman and Danielson (1953) of the formation of globulins in pea roots, indicate that the formation of oligopeptides is a limiting factor in protein synthesis.

It consequently appears that protein synthesis in a cell is simultaneously limited by two factors: the mitochondrial mechanism whereby peptide chains are synthesized and the synthesis of ribonucleoproteins which are necessary for the formation of microsomes, these, in turn, giving proteins their final form by transpeptidation.

It is difficult to understand such a double limitation of the protein-synthesizing mechanism, unless we suppose that both limiting factors can be referred to a common factor. In other words, we must seek either a factor in the mitochondrial mechanism which determines both the peptide synthesis and the formation of RNA in the cell, or a mechanism whereby the synthesis of RNA can determine the activity of the mitochondria in the formation of peptides.

Let us first consider the second alternative. It is generally accepted that the synthesis of ribonucleoproteins is intimately related with the nucleus, and especially with the nucleolus.

It is not, however, quite clear whether the contribution of the nucleolus. which is situated near the heterochromatin and consists largely of histones and small amounts of RNA, is to give up these histones which induce the synthesis of RNA outside the nuclear membrane (cf. CASPERSSON 1950); or whether RNA is formed at the same time as the nucleolus, which effects its transport from the nucleus to the cytoplasm; or, finally, whether the nucleolus contains a precursor of RNA, which acquires its final form in the cytoplasm. The second alternative is preferable to the first, in view of the fact that, in experiments with P^{32}, nucleolar RNA shows a higher specific activity than the RNA in the cytoplasm (BERGSTRAND, ELIASON, HAMMARSTEN, NORBERG, REICHARD and von UBISCH 1948, MARSHAK 1948, JEENER and SZAFARZ 1950, BARNUM and HUSEBY 1950). (See Fig. 30 B.) The third alternative, which may be regarded as a modification of the second, has been made necessary by the finding that nucleolar and cytoplasmic RNA differ in adenine content (MARSHAK 1951, ELSON and CHARGAFF 1951).

Departing from the fact that the synthesis of RNA is under the control of the nucleus, BRACHET (1952 a) has studied the importance of RNA synthesis for protein synthesis by comparing different metabolic data for fragments of amoebae with and without the nucleus. He found that the content of RNA in the enucleated fragment decreased steadily, reaching after 10 days a value about one third of that in the control. The protein content fell off during the same period, though not quite in parallel with the RNA. A further effect of the enucleation was that the uptake of P^{32} from the medium rapidly decreased, while the respiration remained at the level of the control. Since this effect made its appearance almost immediately, much earlier than the decline in protein and RNA. BRACHET concludes that the primary function of the nucleus in this system is to render the mitochondria capable of oxidative phosphorylation. Since the RNA produced by the nucleus is richer in adenine than the corresponding cytoplasmic material, and since the nucleus is known to contain enzymes engaged in the synthesis of nucleotides (STERN, ALLFREY, MIRSKY and SAETREN 1952, HOGEBOOM. SCHNEIDER and STRIEBICH 1952), BRACHET considers that the nucleus is responsible, simultaneously with its production of RNA, for providing the mitochondria with the nucleotides necessary for oxidative phosphorylation. In this view, therefore, the common factor for the synthesis of RNA and for the endergonic peptide synthesis in the mitochondria will derive from the nucleus.

It is hardly possible, however, to accept unreservedly Brachet's hypothesis of a direct nuclear control over oxidative phosphorylation in the mitochondria. Brachet (1952 b) himself has recently found reason to doubt its validity in the course of a study of the activities of certain phosphorylating enzymes in non-nucleated cell fragments. In our opinion the difficulty in accepting Brachet's hypothesis lies in the fact that he has not satisfactorily established that the cessation of uptake of P^{32} in his non-nucleated fragments was actually due to a failure of oxidative phosphorylation, and not to a reduced power of the material to take up phosphate. The latter possibility can hardly be excluded when permeability to neutral red is used as control. The effect of enucleation is scarcely comparable, as Brachet (1952 a) claims, with a dinitrophenol effect, since this substance is known to inhibit the synthetic activity *parallel* with the reduction in phosphate uptake, while enucleation *immediately* stops phosphate uptake and shows only a *gradual* effect on synthetic activity.

There is obviously no doubt that the nucleus has a fundamental influence on oxidative phosphorylation in the mitochondria, and it is highly probable that this influence is associated with the participation of the nucleus in the synthesis of nucleotides. It nevertheless appears that we must seek a *primary* effect in the opposite direction, *i. e.* an effect of the mitochondria on the nucleus. The synthesis of nucleotides and RNA, just as that of peptides and, indeed, synthetic activity in general, is immediately dependent upon energy-generating processes. Regarding the synthesis of RNA, it has been emphasized by several authors (Claude 1948, Zamecnik 1952) that a suitable balance between aerobic and anaerobic processes is a matter of primary importance, in order that an adequate supply of C_2 and C_3 units may be made available for the synthesis of nucleotides. It therefore seems to us that the limitation of protein synthesis may be located, according to the *first* of the alternatives listed above, in a factor in the mitochondrial mechanism which has the double rôle of determining protein synthesis and the formation of ribonucleic acid.

This common factor must constitute a restriction in the aerobic respiratory system and will thus regulate the production of active carbon compounds for the synthesis of both peptides and RNA (see earlier discussion of the regulatory mechanism in mitochondrial activity). The two types of synthesis will hence compete for building units and for energy, while the final elaboration of proteins will be dependent upon RNA. Such a constellation where one process competes with another at *one* point and furthers it at a *second* point, may provide an explanation of the fact that in certain growth-inhibited systems, the usual parallel between the syntheses of protein and RNA is absent (cf. Jeener 1952).

2. Reproduction of cytoplasmic particles

The participation of the microsomes in protein synthesis brings the problem of their reproductive cycle into prominence, since it implies that they may play a part in the cell of great genetic importance. CLAUDE (1943 b), and many after him, have made comparisons of microsomes and viruses in terms of their magnitude and chemical composition. This comparison has given rise to the thought that microsomes might be capable of self-duplication and thus possess genetic continuity.

This analogy, although very striking in itself, contributes nothing to an understanding of this reproduction, for the simple reason that we know almost nothing of how viruses reproduce themselves. What little knowledge we have allows us to presume that the reproduction of microsomes cannot be an independent, isolated phenomenon, such as, for example, the division of a bacterium in a nutrient medium, but must be intimately bound up with the cell metabolism and identical with the life cycle of the microsomes in the cell.

a) Genesis of microsomes

According to our present knowledge, the genesis of microsomes appears to take place in the following stages. Ribonucleoproteins are formed within or without the nucleus, at all events with the active participation of the nucleolus (see discussion above). When vigorous synthesis of ribonucleoproteins is taking place, they are first observed as a ring around the nucleus (CASPERSSON cf. 1950, OPIE and LAVIN 1946) and are probably not bound to particles at this stage (JEENER and SZAFARZ 1950).

Little is known about the mechanism of fixation of the ribonucleoproteins to particles. The Golgi region has recently attracted increasing interest as the possible site of this formation of particles. Cytologists are more and more inclined to regard the Golgi apparatus not as a structural entity, but as an assembly area for different kinds of secretions, especially lipid droplets (cf. BOURNE 1950, MARSHALL 1952). These lipids are believed to be produced by mitochondria (HIRSCH 1939, cf. also BOURNE 1950). ZOLLINGER (1950), after consideration of these findings, has suggested that the microsomes are formed in the Golgi region by combination of ribonucleoproteins and lipids.

The microsomes then enter into the service of protein synthesis. Their rôle, which has been discussed at an earlier stage, may be conceived as consisting in a transformation of the protein, perhaps also in the lengthening of the polypeptide chains. It cannot yet be said whether the microsomes are responsible for a net synthesis of protein. It is also unclear whether the microsomes act by binding a given protein in order to bring about a certain transformation, and then release it in readiness for further activity, or whether the activity involves a dissolution of the microsome, so that a new particle must be provided for each catalytic event. As long as these questions remain unanswered, it will be difficult to decide whether there is differentiation among microsomes. It may nevertheless be considered *a priori* that the microsome population in a cell can hardly be homogeneous, in view of the wide diversity of reactions that must be associated with the synthesis of specific proteins.

b) Genesis of mitochondria

In addition to the alternatives considered above, we must also take into account the possibility that the microsomes may retain the synthesized

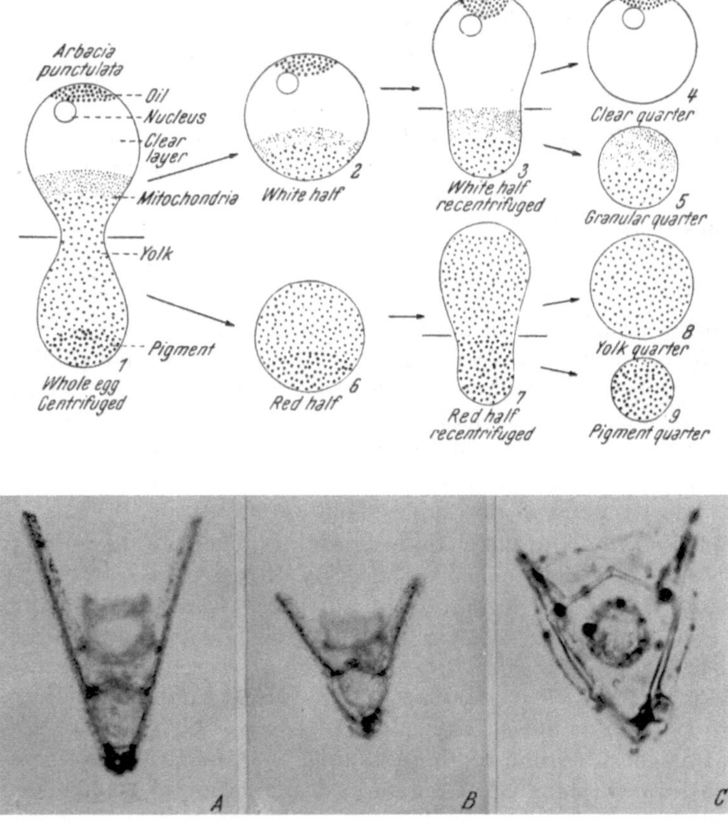

Fig. 31.   Development of larvae (plutei) from egg fragments of the sea urchin, *Arbacia punctulata*, deprived of mitochondria. (By courtesy of Dr. E. B. Harvey.)

Upper figure: Fragmentation of egg by centrifugation (3 min., 10,000 g) (Ann. N.Y. Acad. Sci. **51**, 1337 [1951]). *Lower figures:* Pluteus developed from fertilized A) normal egg (2 days old); B) "white half" (4 days); C) "clear quarter" (10 days). The "clear quarter" pluteus is somewhat retarded, but otherwise relatively normal (J. Exp. Zool. **102**, 271 [1946]).

proteins and gradually be converted to mitochondria (for review, see Runnström 1952). This concept has received increasing experimental support in recent years and has gradually superseded the view that the mitochondria are reproduced by division (Meves 1908). In 1947 (cf. Brachet 1947), Shaver observed that small particles, consisting largely of ribonucleoprotein, were progressively transformed into larger granules

during the early embryonic development of the frog. Similar findings
have been made by LENICQUE (1953). Recently HULTIN (1953), by following
the incorporation of a great variety of isotopically labelled precursors into
the proteins, lipids, nucleic acids and carbohydrates of the different cell
fractions of sea urchin larvae, has been able to confirm and extensively
elucidate this process.

HARVEY (1946), using the centrifuge microscope, has studied the development of fragments of eggs deprived of their mitochondria by centrifugation before fertilization. In these larvae, where, in the view of
RUNNSTRÖM (1952), the surface layer may replace the mitochondria in their
responsibility for cell respiration, development occurs. This is initially
retarded, but is normal in other respects. In the pluteus larva ultimately
formed, there is a normal complement of mitochondria, which shows that
mitochondria have been formed *de novo* in the course of development
(Fig. 31).

ZOLLINGER (1950), in his study of mitochondriogenesis, has made an
approach similar in principle to that outlined above. By repeated intra-
peritoneal injections of egg albumin into mice, he artificially degraded
the mitochondrial population in the proximal convoluted tubules of the
kidney. He was then able to follow the appearance of a new population
in these cells. This occurred as a successive development of small
particles, initially unable to act as mitochondria, through a series of stages
to fully developed mitochondria, capable again of absorbing albumin.
A detailed study of the intermediate stages in the development has recently
been made by EICHENBERGER (1953) (see also p. 10). He followed the
process with the aid of electron micrographs which showed that the
particles first became enlarged by the deposition of protein. A membrane
was then formed, after which the particles could act as mitochondria.

c) The problem of self-duplication of cytoplasmic particles

In the last decades experimental evidence has accumulated for the
existence of genetic continuity in the cytoplasm [3]. There has been
a marked tendency to regard particles as the bearers of this continuity
and hence to attribute to them powers of self-duplication. In certain
cases, it has been claimed that the self-reproductive power has been
demonstrated. These include the plastids of plants, the kinetosomes
studied by LWOFF and the kappa particles of SONNEBORN.

Among the foremost advocates of the self-duplication of the cytoplasmic particles are BRACHET and his collaborators, who have repeatedly
presented experimental evidence in its favour. In one of their principal
investigations (SHAVER and BRACHET 1949), they have attempted to show
that suspensions of microsomes injected into dividing frog eggs, or cultiv-

---

[3] For literature, see: *Colloques Internationaux du Centre National de la Recherche Scientifique*, vol. **VIII**, Paris 1949, and *Cold Spring Harbor Symposia
Quant. Biol.* **18** (1951).

ated on the chorio-allantoic membrane of the chick embryo, can reproduce themselves. Owing to the delicacy of the technique, however, no conclusive results have been obtained (cf. Brachet 1952 a).

A recent paper by Jeener (1952) may be regarded as constituting a further stage in this investigation. This author studied the turnover of RNA in cultures of the flagellate *Polytomella coeca*.

The cultivation was first carried out in a phosphate-poor medium containing P^{32}, until there was no further increase in growth. The RNA content of the cells had then fallen to about one tenth of the initial value. An excess of unlabelled phosphate was then added to the medium, whereupon both the RNA and the protein contents rapidly decreased and after a few hours suddenly began to rise again. Increase in RNA and growth then continued over a period of 24 hours. Analysis of the cell material after centrifugation at 40,000 g showed that the abrupt decrease in the RNA content after addition of excess phosphate was confined to the supernatant fraction, the RNA in the pellet remaining constant. The specific activity, on the other hand, sank uniformly in both fractions, irrespective of their RNA contents.

From these data, Jeener drew the conclusion that the rate of turnover of RNA in a given cell fraction at any moment was a function of the RNA content of the fraction at that time. He demonstrated that the RNA contents in a number of particulate fractions of different sedimentation rates, included in the aforementioned pellet, each showed characteristic variation in the course of the cultivation. He concluded that the cytoplasm contained an indeterminate number of types of particle, each of which was autonomous in its reproduction and independent of the other cell constituents for its life cycle. If this were the case, mitochondria, for example, could not derive from microsomes. The wide bearing of these conclusions makes necessary, however, a more thorough analysis of the experimental data.

Jeener's conclusion is based on the postulate that the uniform decrease in the specific activity of the RNA phosphate in the two cell fractions, in one of which the RNA fell off while it remained constant in the other, must imply that the turnover rate of RNA phosphate is a function of the amount of RNA present in the fraction. As an alternative explanation for the phenomenon, the author considers the possibility that "the rate of renewal of RNA molecules or of the phosphate groups of their constituent nucleotides is sufficiently high so that the specific radioactivity of the RNA remains always independent of quantitative variations which it undergoes." In order to exclude this alternative, Jeener has determined the rate of renewal of RNA in the *Polytomella* culture during the growth period, when the RNA content per cell is constant, and found that this rate is quite low, 3 to 6% per hour of the RNA content.

In our opinion, however, this value does not hold for periods when the RNA content per cell is not constant, and especially not for the period immediately after the addition of phosphate in excess, when it is definitely known that the turnover of RNA, both its *synthesis* and *degradation*, must be appreciably higher than during the period of constancy. The *degradation* is, indeed, initially very vigorous as a result, as Jeener himself suggests, of the large increase in respiration following the addition of phosphate. Similarly, the *synthesis* of RNA must be very lively during this initial period if the content per cell is to increase tenfold, as also reported by the author

It thus seems that JEENER's hypothesis has not received adequate confirmation. Especially difficult to accept is the suggestion that the life cycle of the mitochondria is independent of the microsomes. Even if we disregard the important contributions made earlier by the Belgian school (BRACHET 1947) in the elucidation of the morphogenetic relationship between microsomes and mitochondria, the above-mentioned results of HARVEY and ZOLLINGER are quite decidedly in conflict with a self-duplication of mitochondria. Thus, HARVEY observed the appearance of a mitochondrial population in a system initially devoid of mitochondria; ZOLLINGER deprived a cell of its mitochondria by artificial degeneration and found that the more this degeneration proceeded, the more was the cell stimulated to develop a new mitochondrial population, precisely the reverse of what would be expected if self-duplication occurred.

The fundamental question is whether self-duplication of particles is an indispensable condition of genetic continuity. In other words, is not a self-duplication of particles merely a primitive mode of reproduction which, at a higher level of development, is replaced by a cyclic process, based on a reciprocal action of different constituents in an integrated system (cf. DELBRÜCK 1949, LWOFF 1950)? This query is especially apposite in relation to genetic continuity in the cytoplasm. Here we can observe extensive reciprocal action between different constituents, although we cannot point out any one of these which is responsible for the continuity by virtue of a power of self-duplication. We may indeed ask if we are not seeking the starting point of a circle.

d) Some aspects of the plasmagene problem

The maintenance of genetic continuity in the cytoplasm should be sought in processes which take place entirely in the cytoplasm (cf. MONOD 1949). *Plasmagene products* must thus be substances which are formed in the cytoplasm and contribute to the formation of cytoplasmic elements which further the mechanism whereby the *plasmagene products* are formed (cf. RUNNSTRÖM 1952). As we have already seen, the formation of a microsome consists in the extension of a nuclear product, *i. e.* ribonucleoproteins, with cytoplasmic material, *i. e.* lipids. Microsomes can be developed to mitochondria, while the mitochondria can produce the lipids contributed by the cytoplasm for the development of microsomes.

We are highly ignorant of the biochemistry and metabolism of the lipids. Nevertheless, the importance of these materials in processes allied with genetic phenomena has been repeatedly emphasized in recent years. We need only refer here to the classical investigations by MOEWUS of *Chlamydomonas* (for review, see SONNEBORN 1951), the studies by KOSTERLITZ (1947) and by DAVIDSON and LESLIE (1950), indicating an intimate relationship between protein and lipid metabolism (cf. VENDRELY 1950) and, finally, the important changes in lipid metabolism following carcinogenesis, a matter recently reviewed by Zamecnik (1952). It is perhaps not un-justified to suggest that an increased interest in the biochemistry of the lipids, together with the attention already devoted to the nucleoproteins, will do much to solve the plasmagene problem.

Summary

An important part of the cytoplasm, up to 50% of its total mass, con-sists of particulate elements. Some of these, the mitochondria, are visible under the microscope and are usually rod-shaped, while others, the micro-somes, are of submicroscopic dimensions.

The cytoplasmic particles are not random conglomerates of amorphous material, but well-defined constituents of all living cells. Their number and orientation within a cell may be typical of its metabolic state. The structure of the mitochondria has been fairly well established mor-phologically and physicochemically. They consist of a solid core and a liquid phase, the whole being surrounded by a selectively permeable lipoprotein membrane. Both types of particle are characterized by a high content of lipids: the mitochondria contain all the iron-porphyrin respiratory enzymes of the cell, while the microsomes are bearers of the greater part of the cytoplasmic nucleic acid.

A detailed and systematic study of the enzymic organization of the mitochondria has led in recent years to the belief that these particles contain the apparatus used by the cell for the liberation of the energy of substrates and its conversion into forms suitable for anabolic processes and the performance of work. Energy is liberated in the oxidative pro-cesses of the Krebs cycle, which constitutes a common furnace for the combustion of split products of carbohydrates, fats and proteins. The sphere of activity of the mitochondria extends from pyruvate, fatty acids and transaminatable amino acids, to carbon dioxide and water, the pro-ducts of their total oxidation by way of the Krebs cycle. Furthermore, the mitochondria contain the entire mechanism whereby the energy liberated in these oxidations is fixed in the so-called energy-rich bonds The energy from the primary substances containing such bonds is transferred reversibly by the mitochondrial system to ATP. This energy-transferring system has the function of "activating" the substances which are to participate in various synthetic processes.

By the concentration of its complement of respiratory and energy-transferring enzymes and coenzymes in the mitochondrial organization, the cell achieves the maximum capacity of oxidation and energy transfer attainable with this complement. Since the oxidation of the different substrates, while resulting in the generation of energy, also involves the destruction of building elements necessary for certain syntheses, the high oxidative capacity of the mitochondria is subject to extensive control by the factors responsible for maintaining a balance between oxidative generation of energy and the supply of synthetic material. The mitochondria of tumours, for example. are characterized by a low respiratory power, which may be associated with a deficient mitochondrial organization.

Since the physiological limitation of cell respiration appears to lie in the formation of ATP, it would seem that the factors responsible for the splitting of ATP with the formation of phosphate esters, are also responsible for the aforementioned control mechanism. The enzyme glucokinase, which is known to be subject to hormonal control in the intact cell structure, is largely bound to cytoplasmic particles.

The activation processes consist in the production by the mitochondria of precursors to the substances synthesized by the cell. Thus, apart from the enzymic spectrum necessary for oxidative processes and the formation of ATP, the mitochondria also contain specific activation mechanisms, the nature of which is determined by the type of cell in which they are contained. Specialization can be observed in the sarcosomes of muscle tissue, which serve to provide the contractile elements with ATP: in the mitochondria of kidney tubules, which are concerned in the processes of reabsorption; in liver mitochondria, which specialize in the production of acetoacetate and in the synthesis of citrulline and detoxication conjugates of aromatic compounds: and not least in the mitochondria of embryonic tissue which, in accordance with the prevailing stage of development and the part of the embryo in which they are situated, play an active part in the differentiation process.

A peculiar form of mitochondrial specialization is found in the mechanism of formation of cell secretions and cell inclusions. It is believed that all cell activity of this kind is a typical mitochondrial function. The products are formed by, and accumulated in, the mitochondria. They progressively suppress the mitochondrial structure and the particle, having suffered a great increase in volume, is converted into a droplet of secretion or inclusion.

The RNA-rich submicroscopic particles of the cytoplasm are believed to be mainly concerned in protein metabolism. The intensity of protein synthesis in a cell is known to be correlated with the RNA content of the cytoplasm. Isotope experiments *in vivo* show that the renewal of RNA is most rapid in the nucleus, while the turnover of proteins is most vigorous in the microsomes. The microsomes are therefore supposed to be the organs in which the protein-synthesizing activity of the nucleus in brought to completion. The ribonucleoproteins deriving from the nucleolus are probably the active principles for the protein metabolism of the microsomes.

The mechanism of this activity presumably consists in transpeptidation reactions carried out on protein frameworks conjugated with the particles. The energy for this process is provided by the mitochondria. perhaps in the form of certain "key peptides" containing γ-glutamyl groups.

The study of cytoplasmic particles gives insight into the reciprocal physiological action between the nucleus and the cytoplasm. Our present knowledge of this action is summarized tentatively in Fig. 52.

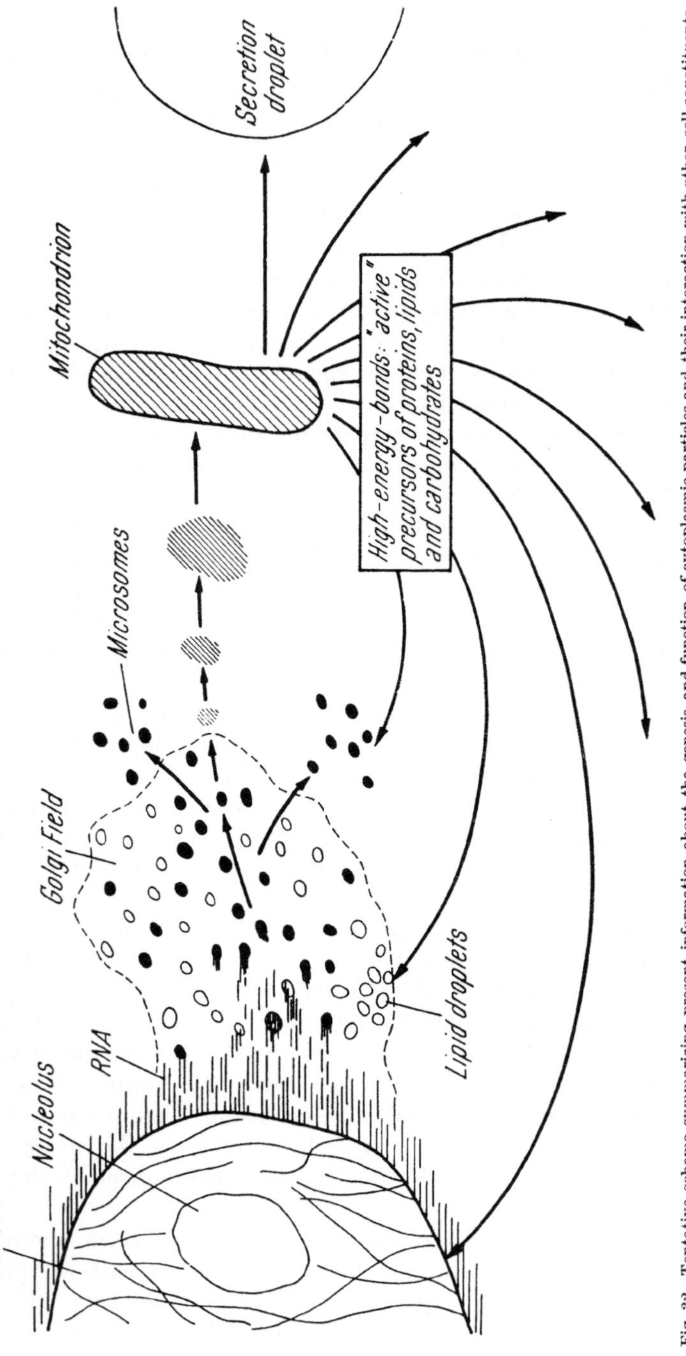

Fig. 32. Tentative scheme summarizing present information about the genesis and function of cytoplasmic particles and their interaction with other cell constituents.

The diagram is intended to show how, on the basis of available information concerning the physiology of the cytoplasmic particles, we are led to conceive the part played by the cytoplasm in the maintenance of the life of the cell, the determination of its differentiation and the preservation of its genetic continuity.

The microsome is a combination of ribonucleoprotein, synthesized under the control of the nucleus, with lipids of the cytoplasm. These latter are believed to be mitochondrial products, stored in the Golgi region as drop-like inclusions.

The microsomes, by virtue of their catalytic power in the re-shaping of proteins, are in all probability involved in the production of the specific proteins of the cell, including the enzymes. There is now evidence from a variety of sources that the mitochondria derive from microsomes by the successive elaboration of a protein structure.

In view of their key function in the transformation of the energy of substrates into forms suited to the special requirements of the cell, the mitochondria may be regarded as playing a decisive part in shaping the metabolic character of the cell within the genetic framework determined by the nucleus. Their contribution to the formation of microsomes by the production of specific lipids would explain, on the basis of a cyclic process, the part played by the cytoplasm in the maintenance of genetic continuity in cell differentiation.

Acknowledgments

We wish to express our cordial thanks to Professor JOHN RUNNSTRÖM for his unstinting interest in our problems during the writing of this monograph and for the inspiring suggestions which he has made during our frequent discussions.

Our gratitude is also due to our colleagues, Dr. T. GUSTAFSON, Dr. T. HULTIN and Miss MARGARETA LJUNGGREN for their valuable help in many matters on which we have sought their guidance.

We thank Drs. C. P. BARNUM, M. EICHENBERGER, G. GLIMSTEDT, T. GUSTAFSON, E. B. HARVEY, T. HULTIN, R. A. HUSEBY, S. LAGERSTEDT, J. RAAFLAUB, J. RHODIN, F. S. SJÖSTRAND, M. I. WATANABE, and C. M. WILLIAMS for allowing us to include microphotographs and diagrams taken from their publications, and the Editors of The Annals of the New York Academy of Science, Archives of Biochemistry, Experimental Cell Research, Helvetica Physiologica Acta, Journal of Experimental Zoology, Journal of General Physiology, and Lunds Universitets Årsskrift for permission to reproduce these pictures.

Drs. C. ALLARD, A. D. BARTON, A. CANTERO, K. W. CLELAND, R. K. CRANE, C. E. DANIELSON, M. EICHENBERGER, P. HAZELTINE, T. HULTIN, A. K. LAIRD, G. DE LAMIRANDE, P. LENICQUE, O. NYGAARD, H. RIS, D. RITCHIE, J. RHODIN, F. S. SJÖSTRAND, O. SNELLMAN, and A. SOLS have very generously allowed us to quote from certain of their papers which have not yet appeared in print.

Finally, we acknowledge our debt to Miss K. PORNER and Mrs. B.-L. WERNER for their help in the preparation of the manuscript.

Translated from the Swedish by

D. F. CHEESMAN, Ph.D. (London).

Submitted for publication, May 16th, 1953.

References

Ada. G. L.. 1949: Phospholipin Metabolism in Rabbit-liver Cytoplasm. Biochem. J. **45**, 422—428.

Ajl. S. J., and C. H. Werkman, 1948: Enzymic Fixation of Carbon Dioxide in α-Ketoglutaric Acid. Proc. Natl. Acad. Sci. **34**, 491—498.

Albaum. H., 1949: Studies.on the Cyclophorase System. XIII. Distribution of Radioactivity in Various Phosphorus-Containing Compounds of the Cyclophorase Gel. Arch. Biochem. **24**, 375—382.

Allard, C., G. de Lamirande, and A. Cantero, 1952: Mitochondrial Population of Mammalian Cells. II. Variation in the Mitochondrial Population of the Average Rat Liver Cell During Regeneration. Use of the Mitochondria as a Unit of Measurement. Cancer Research **12**, 580—585.

— — — 1953: Mitochondrial Population of Mammalian Cells. III. Number of Mitochondria per Average Cell of Rat Liver Tumor Induced by p-Dimethylaminoazobenzene. Significance in the Comparative Study of the Mitochondrial Fraction Properties of Normal Tissues and Tumors. Canad. J. med. Sci. **30**, 543—548.

— R. Mathieu, G. de Lamirande, and A. Cantero, 1952: Mitochondrial Population of Mammalian Cells. I. Description of a Counting Technic and Preliminary Results on Rat Liver in Different Physiological and Pathological Conditions. Cancer Research **12**, 407—412.

Altmann. R., 1890: Die Elementarorganismen und ihre Beziehungen zu den Zellen. Leipzig.

Baddiley. J.. and E. M. Thain. 1951: Coenzyme A. Part II. Evidence for its Formulation as a Derivative of Pantothenic Acid-4-Phosphate. J. Chem. Soc. Part III. 2253—2258.

— — G. D. Novelli. and F. Lipmann. 1953: Structure of Coenzyme A. Nature **171**, 76.

Barker. H. A., 1951: Recent Investigations on the Formation and Utilization of Active Acetate. Phosphorus Metabolism **1**, 204—243.

Barkulis, S. S.. and A. L. Lehninger, 1951: Myokinase and the Adenine Nucleotide Specificity in Oxidative Phosphorylations. J. biol. Chem. (Am.) **190**, 339—344.

Barnum, C. P., and R. A. Huseby. 1948: Some Quantitative Analyses of the Particulate Fractions from Mouse Liver Cytoplasm. Arch. Biochem. **19**, 17—25.

— — 1950: The Intracellular Heterogeneity of Pentose Nucleic Acid as Evidenced by the Incorporation of Radiophosphorus. Arch. Biochem. **29**, 7—26.

Bartley, W., 1953: An Effect of Bicarbonate on the Oxidation of Pyruvate by Kidney Homogenates. Biochem. J. **53**, 305—312.

Beinert. H., R. W. von Korff, D. E. Green, D. A. Buyrke. R. E. Handschumacher, H. Higgins, and F. M. Shong. 1952: A Method for Purification of Coenzyme A. J. amer. chem. Soc. **74**, 854—855.

Bensley, S. H., 1947: The Normal Mode of Secretion in the Parathyroid Gland of the Dog. Anat. Record **98**, 361—377.

Bensley. R. R., and S. H. Bensley, 1938: Handbook of Histological and Cytological Technique. Chicago.

— and N. Hoerr, 1934: The Preparation and Properties of Mitochondria. Anat. Record **60**, 449—455.

Bergstrand. A.. N. A. Eliason, E. Hammarsten, B. Norberg, P. Reichard, and H. von Ubisch, 1948: Experiments with N^{15} on Purines from Nuclei and Cytoplasm of Normal and Regenerating Liver. Cold Spring Harbor Symposia Quant. Biol. **13**, 22—25.

Bernheim. F.. K. M. Wilbur. and C. B. Kenaston. 1952: The Effect of Oxidized Fatty Acids on the Activity of Certain Oxidative Enzymes. Arch. Biochem. and Biophys. **38**, 177—184.

Bernstein, S.. and R. W. McGilvery, 1952: Substrate Activation in the Synthesis of Phenyl Sulfate. J. biol. Chem. (Am.) **199**, 745—748.

Berthet, J.. L. Berthet, F. Appelmans, and C. de Duve, 1952: Tissue Fractionation Studies. II. The Nature of the Linkage between Acid Phosphatase and Mitochondria in Rat Liver Tissue. Biochem. J. **50**, 182—189.

— and C. de Duve. 1952: Tissue Fractionation Studies. I. The Existence of a Mitochondria-Linked. Enzymatically Inactive Form of Acid Phosphatase in Rat Liver Tissue. Biochem. J. **50**, 174—181.

Beyer, K. H., L. D. Wright, H. R. Skeggs, H. F. Russo, and G. A. Shaner, 1947: Renal Clearance of Essential Amino Acids: Their Competition for Reabsorption by the Renal Tubules. Amer. J. Physiol. **151**, 202—210.

Binkley, F., 1952: Evidence for the Polynucleotide Nature of Cysteinylglycinase. Exptl. Cell Research **Suppl. 2**, 145—160.

Borsook, H., C. L. Deasy, A. J. Haagen-Smit, G. Keighley, and P. H. Lowy, 1949: The Incorporation of Labelled Lysine into the Proteins of Guinea Pig Liver Homogenate. J. biol. Chem. (Am.) **179**, 689—719.

— — — — — 1950: Incorporation of C^{14}-Labelled Amino Acids into Proteins of Fractions of Guinea Pig Liver Homogenate. Federation Proc. **9**, 154—155.

— and J. W. Dubnoff, 1941: The Conversion of Citrulline to Arginine in Kidney. J. Biol. Chem. **141**, 717—738.

Bourne, G., 1942: Cytology and Cell Physiology. London.

Bourne, G. H., 1950: XX. Recent Discoveries Concerning Mitochondria and Golgi Apparatus and their Significance in Cellular Physiology. J. Roy. Microscop. Soc. **70**, 367—380.

Brachet, J., 1940: La détection histochimique des acides pentosenucléiques. C. r. Soc. Biol. **133**, 88—90.

— 1941: La détection histochimique et le microdosage des acides pentosenucléiques. (Tissus animaux — développement embryonnaire des amphibiens.) Enzymologia **X**, 87—96.

— 1947: The Metabolism of Nucleic Acids During Embryonic Development. Cold Spring Harbor Symposia Quant. Biol. **12**, 18—27.

— 1952 a: The Rôle of the Nucleus and the Cytoplasm in Synthesis and Morphogenesis. Symposia Soc. Exptl. Biol. **6**, 173—200.

— 1952 b: Le rôle du noyau cellulaire dans les oxydations et les phosphorylations. Biochim. et Biophys. Acta **9**, 221—222.

Brady, R. O., and S. Gurin, 1952: Biosynthesis of Fatty Acids by Cell-free or Water-Soluble Enzyme Systems. J. biol. Chem. (Am.) **199**, 421—431.

Breusch, F. L., in J. B. Sumner, and K. Myrbäck, 1952: Carbohydrate-Fat Conversion. Enzymes **2**, 1035—1051.

Brody, T. M., and J. A. Bain, 1952: A Mitochondrial Preparation from Mammalian Brain. J. biol. Chem. (Am.) **195**, 685—696.

— R. J. H. Wang, and J. A. Bain, 1952: Intracellular Distribution of Diphosphopyridine Nucleotide-Cytochrome c Reductase and Cytochrome c Oxidase in Mammalian Brain. J. biol. Chem. (Am.) **198**, 821—826.

Bullough, W. S., 1952: The Energy Relations of Mitotic Activity. Biol. Rev. Cambridge philos. Soc. **27**, 133—168.

— 1955: in Ciba Foundation Colloquia on Endocrinology **6**, 278—289.

Cantoni, G. L., 1951: On the Rôle of High Energy Phosphate in Transmethylation. Phosphorus Metabolism **1**, 641—646.

— 1952: The Nature of the Active Methyldonor Formed Enzymatically from l-Methionine and ATP. J. amer. chem. Soc. **74**, 2942—2943.

Caputto, R., 1951: The Enzymatic Synthesis of Adenylic Acid; Adenosinekinase. J. biol. Chem. (Am.) **189**, 801—814.

Caspersson, T., 1941: Studien über den Eiweißumsatz der Zelle. Naturw. **29**, 33—43.

— 1947: The Relations between Nucleic Acid and Protein Synthesis. Symposia Soc. Exptl. Biol. **1**, 127—151.

— 1950: Cell Growth and Cell Function. New York.

Chance, B., and L. Smith, 1952: Biological Oxidations. Ann. Rev. Biochem. **21**, 687—726.

Chantrenne, H., 1947: Hétérogénéité des granules cytoplasmiques du foie de souris. Biochim. et Biophys. Acta **1**, 437—448.

— 1951: The Requirement for Coenzyme A in the Enzymatic Synthesis of Hippuric Acid. J. biol. Chem. (Am.) **189**, 227—233.

Chou, T. C., and F. Lipmann, 1952: Separation of Acetyl Transfer Enzymes in Pigeon Liver Extract. J. biol. Chem. (Am.) **196**, 89—103.

— G. D. Novelli, E. R. Stadtman, and F. Lipmann, 1950: Fractionation of Coenzyme A-Dependent Acetyl Transfer Systems. Federation Proc. **9**, 160.

— and M. Soodak, 1952: The Acetylation of d-Glucosamine by Pigeon Liver Extract. J. biol. Chem. (Am.) **196**, 105—109.

CLAUDE, A., 1940: Particulate Components of Normal and Tumor Cells. Science **91**, 77—78.
— 1943 a: Distribution of Nucleic Acids in the Cell and the Morphological Constitution of Cytoplasm. Biol. Symposia **10**, 111—199.
— 1943 b: The Constitution of Protoplasm. Science **97**, 451—456.
— 1944: The Constitution of Mitochondria and Microsomes and the Distribution of Nucleic Acid in the Cytoplasm of a Leukemic Cell. J. Exptl. Med. **80**, 19—29.
— 1946: Fractionation of Mammalian Liver Cells by Differential Centrifugation. I. Problem, Method, and Preparation of Extract. II. Experimental Procedures and Results. J. exper. Med. (Am.) **84**, 51—89.
— 1948: Studies on Cells: Morphology, Chemical Constitution, and Distribution of Biochemical Functions. Harvey Lectures **Ser. 43**, 121—164.
— and E. F. FULLAM, 1945: An Electron Microscope Study of Isolated Mitochondria. J. exper. Med. (Am.) **81**, 51—62.
— — 1946: The Preparation of Sections of Guinea Pig Liver for Electron Microscopy. J. exper. Med. (Am.) **83**, 499—504.
CLELAND, K. W., 1952: Permeability of Isolated Rat Heart Sarcosomes. Nature **170**, 497—499.
— and SLATER E. C., 1953: Respiratory granules of heart muscle. Biochem. J. **53**, 547—556.
COHEN, Ph. P., 1951: The Utilization of Phosphate Bond Energy in Biological Systems. The Synthesis of Peptide Bonds. Phosphorus Metabolism **1**, 630—640.
— and R. W. McGILVERY, 1947: Peptide Bond Synthesis. III. On the Mechanism of p-Aminohippuric Acid Synthesis. J. biol. Chem. (Am.) **171**, 121—133.
COLOWICK, S. P., 1951 a: The Reduction of DPN by TPNH. Phosphorus Metabolism **1**, 436—442.
— in J. B. SUMNER, and K. MYRBÄCK. 1951 b: Transphosphorylating Enzymes of Fermentation. Enzymes **2**, 114—150.
— G. T. CORI, and M. W. SLEIN, 1947: The Effect of Adrenal Cortex and Anterior Pituitary Extracts and Insulin on the Hexokinase Reaction. J. biol. Chem. (Am.) **168**, 583—596.
— and H. M. KALCKAR, 1943: The Rôle of Myokinase in Transphosphorylations. I. The Enzymatic Phosphorylation of Hexoses by Adenylpyrophosphate. J. biol. Chem. (Am.) **148**, 117.
— N. O. KAPLAN, E. F. NEUFELD, and M. M. CIOTTI, 1952: Pyridine Nucleotide Transhydrogenase. I. Indirect Evidence for the Reaction and Purification of the Enzyme. J. biol. Chem. (Am.) **195**, 95—105.
— M. S. WELCH, and C. F. CORI, 1940: Phosphorylation of Glucose on Kidney Extract. J. biol. Chem. (Am.) **133**, 359—373.
COOPERSTEIN, S. J., and A. LAZAROW, 1953: Studies on the Mechanism of Janus Green B Staining of Mitochondria. III. Reduction of Janus Green B by Isolated Enzyme System. Exptl. Cell Research **5**, 82—97.
— — and I. W. PATTERSON, 1953: Studies on the Mechanism of Janus Green B Staining of Mitochondria. II. Reactions and Properties of Janus Green B and its Derivates. Exptl. Cell Research **5**, 69—82.
CORI, C. F., 1949: Influence of Hormones on Enzymatic Reactions. Report of Opening and Concluding Sessions and Three Lectures. First International Congress of Biochemistry. Cambridge. 9—18.
— S. F. VELICK, and G. T. CORI, 1950: The Combination of Diphosphopyridine Nucleotide with Glyceraldehyde Phosphate Dehydrogenase. Biochim. et Biophys. Acta **4**, 160—169.
— M. W. SLEIN, and C. F. CORI, 1948: Crystalline d-Glyceraldehyde-3-Phosphate Dehydrogenase from Rabbit Muscle. J. biol. Chem. (Am.) **173**, 605—618.
COWDRY, E. V., 1928: Special Cytology. New York.
— 1943: Microscopic Technique in Biology and Medicine. Baltimore.
CRANE, R. K., and A. SOLS, 1953: The Association of Hexokinase with Particulate Fractions of Brain and other Tissue Homogenates. J. biol. Chem. (Am.) **203**, 273—292.
CROSS, R. J., and J. V. TAGGART, 1950: Renal Tubular Transport: Accumulation of p-Aminohippurate by Rabbit Kidney Slices. Amer. J. Physiol. **161**, 181—190.
— J. V. TAGGART, A. G. COVO, and D. E. GREEN, 1949: Studies on the Cyclophorase System. VI. The Coupling of Oxidation and Phosphorylation. J. biol. Chem. (Am.) **177**, 655—678.

DALTON. A. J., H. KAHLER, M. G. KELLY, B. J. LLOYD. and M. J. STRIEBICH, 1949: Some Observations on the Mitochondria of Normal and Neoplastic Cells with the Electron Microscope. J. Natl. Cancer Inst. 9, 439—449.

DARLING, S., 1952: Cysteic Acid Transaminase. Nature 170, 749—750.

DAVIDSON, J. N., and J. LESLIE, 1950: A New Approach in the Biochemistry of Growth and Development. Nature 165, 49—53.

DELBRÜCK, M., 1948: Discussion. VIII. Unités Biologiques Douées de Continuité Génétique. Colloq. Intern. centre natl. recherche sci. Paris. 33—35.

DRYSDALE, G. R., 1952: β-Oxidation of Fatty Acids by a Soluble Enzyme System from Mitochondria. Federation Proc. 11, 204.

DE DUVE, C., F. APPELMANS, and R. WATTIAUX, 1952: Cytochrome-Oxidase and Acid Phosphatase of Isolated Mitochondria. Résumés des Communications. IIe Congrès International de Biochimie. Paris. 278.

— J. BERTHET, L. BERTHET, and F. APPELMANS, 1951: Permeability of Mitochondria. Nature 167, 389—390.

EDELHOCH, H., O. HAYAISHI, and L. J. TEPLY, 1952: The Preparation and Properties of a Soluble Diphosphopyridine Nucleotide Cytochrome c Reductase. J. biol. Chem. (Am.) 197, 97—104.

EICHENBERGER, M., 1953: Elektronenmikroskopische Beobachtungen über die Entstehung der Mitochondrien aus Mikrosomen. Exptl. Cell Research 4, 275—282.

EILER, J. J., and W. K. McEWEN, 1949: The Effect of Pentobarbital on Aerobic Phosphorylation in Brain Homogenates. Arch. Biochem. 20, 163—165.

ELLIOTT, W. H., 1951: Studies on the Enzymic Synthesis of Glutamine. Biochem. J. 49, 106—112.

ELSON, D., and E. CHARGAFF. 1951: Nucleotide Composition of Pentose Nucleic Acids from Different Fractions of the Hepatic Cell. Federation Proc. 10, 180—181.

ERNSTER, L., and O. LINDBERG, 1952: Studies on the Mechanism of Oxidative Phosphorylation in Mitochondria. Résumés des Communications. IIe Congrès International de Biochimie. Paris. 33.

FORSTER, R. P., and J. V. TAGGART, 1950: Use of Isolated Renal Tubules for the Examination of Metabolic Processes Associated with Active Cellular Transport. J. cellul. a. comp. Physiol. (Am.) 36, 251—270.

FRÉDÉRIC, J., et M. CHÈVREMONT, 1952: Recherches sur les chondriosomes de cellules vivantes par la microscopie et la microcinématographie en contraste de phase. Arch. Biol. 63, 109—131.

FREI. J., et F. LEUTHARDT, 1949: La synthèse biologique de la glutamine. Helv. Chim. Acta 32, 1137—1143.

FRIEDKIN. M., and A. L. LEHNINGER, 1947: The Synthesis of Phosphomalic Acid. J. biol. Chem. (Am.) 169, 183—190.

— — 1949 a: Oxidation-Coupled Incorporation of Inorganic Radiophosphate into Phospholipide and Nucleic Acid in a Cell-free System. J. biol. Chem. (Am.) 177, 775—788.

— — 1949 b: Esterification of Inorganic Phosphate Coupled to Electron Transport between Dihydrodiphosphopyridine Nucleotide and Oxygen. J. biol. Chem. (Am.) 178, 611—623.

GELOTTE. B., 1951: A Phosphate-Absorbing Protein from Muscular Tissue. Nature 168, 461.

GERGELY, J., P. HELE, and C. V. RAMAKRISHNAN. 1952: Succinyl and Acetyl Coenzyme A Deacylases. J. biol. Chem. (Am.) 198, 323—334.

GLIMSTEDT, G., and S. LAGERSTEDT. 1953: Observations on the Ultrastructure of Isolated Mitochondria from Normal Rat Liver. Lunds Universitets Årsskrift, N. F., Avd. 2, 49, Nr. 5; Kungl. Fysiografiska Sällskapets Handlingar. N. F., 64. Nr. 5, 1—11.

GREEN. D. E., in J. T. EDSALL. 1951 a: The Cyclophorase System. Enzymes and Enzyme Systems. Cambridge. Mass., 15—46.

— 1951 b: The Cyclophorase Complex of Enzymes. Biol. Rev. 26, 410—433.

— 1951 c: Discussion. Phosphorus Metabolism 1, 657.

— 1952: The Cyclophorase-Mitochondrial System. Symposium sur le Cycle Tricarboxylique. IIe Congrès International de Biochimie. Paris. 1—27.

— W. A. ATCHLEY, J. NORDMANN. and L. J. TEPLY. 1949: Studies on the Cyclophorase System. XII. Incorporation of P^{32}. Arch. Biochem. 24, 359—374.

Green. D. E., H. Beinert. M. Fuld. D. Goldmann, M. H. Paul, and N. K. Sarkar, 1953: Studies on the Cyclophorase System. XXVII. Cyclophorase Activity in a Non-Mitochondrial System of Pig Heart Muscle. Exptl. Cell Research 4, 222—235.
— W. F. Loomis, and V. H. Auerbach. 1948: Studies on the Cyclophorase System. I. The Complete Oxidation of Pyruvic Acid to Carbon Dioxide and Water. J. biol. Chem. (Am.) 172. 389—405.
— — S. R. Dickman, V. H. Auerbach. and B. Noyce, 1947: Cyclophorase System of Kidney for Complete Oxidation of Pyruvic Acid. Abstracts of Papers. 111th Meeting Am. Chem. Soc. Atlantic City. p. 26 B.
Gregory. J. D., and F. Lipmann. 1952: The Nature of the Sulfur in Coenzyme A. J. amer. chem. Soc. 74, 4017—4019.
Grisolia. S., 1951: Mechanism of the Biosynthesis of Citrulline. Phosphorus Metabolism 1, 619—629.
— and Ph. P. Cohen, 1952: The Catalytic Rôle of Carbamyl Glutamate in Citrulline Biosynthesis. J. biol. Chem. (Am.) 198, 561—571.
Guilliermond. A., 1932: La structure de la cellule végétale: Les inclusions du cytoplasme. et en particulier les chondriosomes et les plastes. Protoplasma 16, 291—337.
Gunsalus. I. C., M. I. Dolin, and L. Struglia. 1951: Pyruvic Acid Metabolism. J. biol. Chem. (Am.) 194, 849—857.
— L. Struglia, and D. J. O'Kane. 1951: Pyruvic Acid Metabolism. J. biol. Chem. (Am.) 194. 859—869.
Gustafson. T., 1952: Nitrogen Metabolism, Enzymic Activity. and Mitochondrial Distribution in Relation to Differentiation in the Sea Urchin Egg. Uppsala.
— and I. Hasselberg, 1951: Studies on Enzymes in the Developing Sea Urchin Egg. Exptl. Cell Research 2, 642—672.
— M. B. Hjelte, and I. Hasselberg, 1952: Growth Promoting Factors in Developing Sea Urchin Egg. Exptl. Cell Research 3, 275—281.
— and P. Lenique, 1952: Studies on Mitochondria in the Developing Sea Urchin Egg. Exptl. Cell Research 3, 251—274.
Handler. P., M. L. C. Bernheim. and J. R. Klein. 1941: The Oxidative Demethylation of Sarcosine to Glycine. J. biol. Chem. (Am.) 138, 211—229.
Hanes, C. S., F. J. R. Hird. and F. A. Isherwood. 1950: Synthesis of Peptides in Enzymic Reactions Involving Glutathione. Nature 166, 288—292.
Harman, J. W., 1950 a: Studies on Mitochondria. I. The Association of Cyclophorase with Mitochondria. Exptl. Cell Research 1. 382—393.
— 1950 b: Studies on Mitochondria. II. The Structure of Mitochondria in Relation to Enzymatic Activity. Exptl. Cell Research 1. 394—402.
— and M. Feigelson, 1952 a: Studies on Mitochondria. III. The Relationship of Structure and Function of Mitochondria from Heart Muscle. Exptl. Cell Research 3, 47—58.
— — 1952 b: Studies on Mitochondria. V. The Relationship of Structure and Oxidative Phosphorylation on Mitochondria of Heart Muscle. Exptl. Cell Research 3, 509—525.
— — 1952 c: Studies on Mitochondria. IV. The Cytological Localization of Mitochondria in Heart Muscle. Exptl. Cell Research 3. 58—64.
Harting. J., 1951: Oxidation of Acetaldehyde by Glyceraldehyde-3-Phosphate Dehydrogenase of Rabbit Muscle. Federation Proc. 10. 195.
— and S. F. Velick. 1952: Reactions of Acetyl Phosphate Catalyzed by 3-Phosphoglyceraldehyde Dehydrogenase. Federation Proc. 11. 226.
Hartmann. J. F., 1948: Mitochondria in Nerve Cell Bodies Following Section of Axones. Anat. Record 100, 49—59.
Harvey. E. B., 1937: The Centrifuge Microscope. Microscopical Technique. New York.
— 1946: Structure and Development of the Clear Quarter of the *Arbacia punctulata* Egg. J. Exptl. Zool. 102. 253—271.
— and Th. F. Anderson. 1943: The Spermatozoon and Fertilization Membrane of *Arbacia punctulata* as Shown by the Electron Microscope. Biol. Bull. 85. 151—156.
Hers, H. G., J. Berthet. L. Berthet. et C. de Duve. 1951: Le système hexosephosphatasique. III. Localisation intracellulaire des ferments par centrifugation fractionnée. Bull. Soc. Chim. biol. (Fr.) 33. 21—41.

HERSEY, D. F., and S. J. AJL, 1951 a: Phosphorylation Due to the Oxidation of Succinic Acid by Cell-free Extracts of *Escherichia Coli*. J. gen. Physiol. (Am.) **34**, 295—304.
— — 1951 b: Adenosinetriphosphate Formation in the Oxidation of Succinic Acid by Bacteria. J. biol. Chem. (Am.) **191**. 113—121.
HIRD, F. J. R., and E. V. ROWSELL, 1950: Additional Transmination by Insoluble Particle Preparations of Rat Liver. Nature **166**, 517—518.
HIRSCH, G. C., 1939: Form und Stoffwechsel der Golgikörper. Protoplasma Monographs. Berlin.
HOGEBOOM, G. H., 1949: Cytochemical Studies of Mammalian Tissues. II. The Distribution of Diphosphopyridine Nucleotide-Cytochrome *c* Reductase in Rat Liver Fractions. J. biol. Chem. (Am.) **177**, 847—858.
— A. CLAUDE, and R. D. HOTCHKISS, 1946: The Distribution of Cytochrome Oxidase and Succinoxidase in the Cytoplasm of the Mammalian Liver Cell. J. biol. Chem. (Am.) **165**, 615—629.
— and W. C. SCHNEIDER, 1950 a: Sonic Disintegration of Isolated Liver Mitochondria. Nature **166**, 502—503.
— — 1950 b: Cytochemical Studies of Mammalian Tissues. III. Isocitric Dehydrogenase and Triphosphopyridine Nucleotide-Cytochrome *c* Reductase of Mouse Liver. J. biol. Chem. (Am.) **186**, 417—427.
— — 1950 c: Intracellular Distribution of Enzymes. VIII. The Distribution of Diphosphopyridine-Nucleotide-Cytochrome *c* Reductase in Normal Mouse Liver and Mouse Hepatoma. J. Natl. Cancer Inst. **10**, 983—987.
— — 1951: Proteins of Liver and Hepatoma Mitochondria. Science **113**. 355—358.
— — and G. E. PALADE, 1948: Cytochemical Studies of Mammalian Tissues. I. Isolation of Intact Mitochondria from Rat Liver; Some Biochemical Properties of Mitochondria and Submicroscopic Particulate Material. J. biol. Chem. (Am.) **172**, 619—635.
— — and M. J. STRIEBICH, 1952: Cytochemical Studies. V. On the Isolation and Biochemical Properties of Liver Cell Nuclei. J. biol. Chem. (Am.) **196**. 111—120.
HOLTER, H., 1952: Localization of Enzymes in Cytoplasm. Advances in Enzymol. **13**, 1—20.
HOLZER, H., 1952: Acetyl-Coenzyme A und andere S-Acyl-Verbindungen bei der Energieausnützung in der lebenden Zelle. Angew. Chem. **64**, 248—255.
— und E. HOLZER, 1952: Zum Reaktionsmechanismus der Triose-Phosphatdehydrierung. Z. physiol. Chem. **291**, 67—86.
— und F. LYNEN, 1950: Über den aeroben Phosphatbedarf der Hefe. III. Labil an die Struktur gebundenes Phosphat in lebender Hefe. Ann. Chem. Justus Liebigs **569**, 138—148.
HORECKER, B. L., 1950: Triphosphopyridine Nucleotide-Cytochrome *c* Reductase in Liver. J. biol. Chem. (Am.) **183**, 593—605.
HUENNEKENS, F. M., and D. E. GREEN. 1950: Studies on the Cyclophorase System. X. The Requirement for Pyridine Nucleotide. Arch. Biochem. **27**. 418—427.
— H. R. MAHLER, and J. NORDMANN. 1951: Studies on the Cyclophorase System. XVII. The Occurrence and Properties of an α-Hydroxy Acid Racemase. Arch. Biochem. **30**, 77—89.
HUGHES, A. F. W., and H. B. FELL. 1949: Studies on Abnormal Mitoses Induced in Chick Tissue Cultures by Mustard Gas. Quart. J. Microscop. Sci. **90**. 37—55.
HULTIN, T., 1950: Incorporation in vivo of N^{15}-Labelled Glysine into Liver Fractions of Newly Hatched Chicks. Exptl. Cell Research **1**. 376—381.
— 1953: Studies on the Structural and Metabolic Background of Fertilization and Development. Stockholm.
HUNTER, F. E., 1949: Anaerobic Phosphorylation Due to a Coupled Oxidation-Reduction between α-Ketoglutaric Acid and Oxalacetic Acid. J. biol. Chem. (Am.) **177**. 361—372.
— 1951: Oxidative Phosphorylation During Electron Transport. Phosphorus Metabolism **1**, 297—330.
— and W. S. HIXON, 1949 a: Anaerobic Phosphorylation Due to the Dismutation of α-Ketoglutaric Acid in the Presence of Ammonia. J. biol. Chem. (Am.) **181**. 67—71.
— — 1949 b: Phosphorylation Coupled with the Oxidation of α-Ketoglutaric Acid. J. biol. Chem. (Am.) **181**. 73—79.
— and S. SPECTOR, 1951: Effect of Dinitrophenol on Phosphorylations Coupled with Oxidation of α-Ketoglutarate. Federation Proc. **10**. 201.

Hurvitz. J.. 1952: The Enzymic Phosphorylation of Vitamin B$_6$ Derivatives and their Effects on Tyrosine Decarboxylase. Biochim. et Biophys. Acta 9, 496—498.

Huseby. R. A.. and C. P. Barnum, 1950: Investigation of the Phosphorus Containing Constituents of Centrifugally Prepared Fractions from Mouse Liver Cell Cytoplasm. Arch. Biochem. 26. 187—198.

Hutchens. J. O., M. J. Kopac, and M. E. Krahl, 1942: The Cytochrome Oxidase Content of Centrifugally Separated Fractions on Unfertilized *Arbacia* Eggs. J. cellul. a. comp. Physiol. (Am.) 20. 113—116.

Jagannathan, V., and R. S. Schweet. 1952: Pyruvic Oxidase of Pigeon Breast Muscle. I. Purification and Properties of the Enzyme. J. biol. Chem. (Am.) 196, 551—562.

Jeener, R., 1948: L'hétérogénéité des granules cytoplasmiques : données complémentaires fournies par leur fractionnement en solution saline concentrée. Biochim. et Biophys. Acta 2, 633—641.

— 1952: Studies on the Evolution of Nucleoprotein Fractions of the Cytoplasm During the Growth of a Culture of *Polytomella coeca.* Biochim. et Biophys. Acta 8. 270—282.

— and D. Szafarz, 1950: Relations between the Rate of Renewal and the Intracellular Localisation of Ribonucleic Acid. Arch. Biochem. 26, 54—67.

Johnson, R. B., and W. W. Ackermann, 1953: A Rôle of Nuclei in Oxidative Phosphorylation. J. biol. Chem. (Am.) 200, 263—269.

Johnston, R. B., and K. Bloch. 1951: Enzymatic Synthesis of Glutathione. J. biol. Chem. (Am.) 188, 225—240.

Judah, J. D.. and H. G. Williams-Ashman. 1951: The Inhibition of Oxidative Phosphorylation. Biochem. J. 48, 33—42.

Junqueira, L. C. U.. 1951: Cytological. Cytochemical and Biochemical Observations on Secreting and Resting Salivary Glands. Exptl. Cell Research 2, 327—338.

Kalckar. H. M., 1939: The Nature of Phosphoric Esters Formed in Kidney Extracts. Biochem. J. 33. 631—641.

— 1941: The Nature of Energetic Coupling in Biological Synthesis. Chem. Rev. 28. 71.

Kaplan, N. O.. in J. B. Sumner, and K. Myrbäck. 1951 a: Thermodynamics and Mechanism of the Phosphate Bond. Enzymes 2, 55—113.

— 1951 b: Pyridine Nucleotides and Coupled Phosphorylation. Phosphorus Metabolism 1. 428—436.

— S. P. Colowick. and E. F. Neufeld. 1952: Pyridine Nucleotide Transhydrogenase. II. Direct Evidence for and Mechanism of the Transhydrogenase Reaction. J. biol. Chem. (Am.) 195, 107—119.

— and F. Lipmann. 1948: The Assay and Distribution of Coenzyme A. J. biol. Chem. (Am.) 174. 37—44.

— J. L. Still, and H. R. Mahler, 1951: Studies on the Cyclophorase System. XVIII. The Oxido-Reductions of Glycolysis. Arch. Biochem. and Biophys. 34. 16—25.

Kaufman, S., 1951: Soluble α-Ketoglutaric Dehydrogenase from Heart Muscle and Coupled Phosphorylation. Phosphorus Metabolism 1. 370—373.

Kearney, E. B., and S. Englard, 1951: The Enzymatic Phosphorylation of Riboflavin. J. biol. Chem. (Am.) 193, 821—834.

Keilin. D.. 1929: Cytochrome and Respiratory Enzymes. Proc. roy. Soc.. Lond. 104. 206—252.

Keller, E. B.. 1951: Turnover of Proteins of Cell Fractions of Adult Rat Liver in vivo. Federation Proc. 10, 206.

Kennedy, E. P.. 1952: Synthesis of Phospholipids in Isolated Mitochondria. Federation Proc. 11, 239.

— and A. L. Lehninger, 1948: Intracellular Structures and the Fatty Acid Oxidase System of Rat Liver. J. biol. Chem. (Am.) 172. 847—848.

— — 1949: Oxidation of Fatty Acids and Tricarboxylic Acid Cycle Intermediates by Isolated Rat Liver Mitochondria. J. biol. Chem. (Am.) 179. 957—972.

Kielley. W. W.. and R. K. Kielley, 1951: Myokinase and Adenosinetriphosphatase in Oxidative Phosphorylation. J. biol. Chem. (Am.) 191. 485—500.

Kielley. R. K.. and W. C. Schneider, 1950: Synthesis of p-Aminohippuric Acid by Mitochondria of Mouse Liver Homogenates. J. biol. Chem. (Am.) 185. 869—880.

Korey. S. R.. B. de Braganza. and D. Nachmansohn. 1951: Choline Acetylase. V. Esterifications and Transacetylations. J. biol. Chem. (Am.) 189. 705—715.

Korkes, S., A. del Campillo, I. C. Gunsalus, and S. Ochoa, 1951: Enzymatic Synthesis of Citric Acid. IV. Pyruvate as Acetyl Donor. J. biol. Chem. (Am.) **193**, 721—735.

— — S. R. Korey, J. R. Stern, D. Nachmansohn, and S. Ochoa, 1952: Coupling of Acetyl Donor Systems with Choline Acetylase. J. biol. Chem. (Am.) **198**, 215—220.

Kornberg, A., 1950: Reversible Enzymatic Synthesis of Diphosphopyridine Nucleotide and Inorganic Pyrophosphate. J. biol. Chem. (Am.) **182**, 779—793.

— and O. Lindberg, 1948: Diphosphopyridine Nucleotide Pyrophosphatase. J. biol. Chem. (Am.) **176**, 665—677.

— and W. E. Pricer, 1951: Enzymatic Phosphorylation of Adenosine and 2, 6-Diaminopurrine Riboside. J. biol. Chem. (Am.) **193**, 481—495.

— — 1952: Enzymatic Synthesis of Phosphorus-Containing Lipide. J. amer. chem. Soc. **74**, 1617.

Kosterlitz, H. W., 1947: The Effect of Changes in Dietary Protein on the Composition and Structure on the Liver Cell. J. Physiol. **106**, 194—210.

Kotelnikova, A. V., 1950: Biokhimiya **15**, 371.

Krebs, H. A., 1935: Metabolism of Amino-Acids. IV. The Synthesis of Glutamine from Glutamic Acid and Ammonia, and the Enzyme Hydrolysis of Glutamine in Animal Tissues. Biochem. J. **29**, 1951—1969.

— and Ph. P. Cohen, 1939: Metabolism of α-Ketoglutaric Acid in Animal Tissues. Biochem. J. **33**, 1895—1899.

Krimsky, J., and E. Racker, 1952: Glutathione, a Prosthetic Group of Glyceraldehyde-3-Phosphate Dehydrogenase. J. Biol. Chem. (Am.) **198**, 721—729.

Kriszat, G., 1949: Die Wirkung von Adenosintriphosphat auf Amöben (*Chaos chaos*). Arkiv Zool. **1**, 81—86.

— 1950: Die Wirkung von Adenosintriphosphat und Calcium auf Amöben (*Chaos chaos*). Arkiv Zool. (Ser. 2) **1**, 477—485.

— 1951: Die Wirkung von Ca und ATP bei Vitalfärbungsversuchen mit Amöben (*Chaos chaos*). Arkiv Zool. (Ser. 2) **3**, 115—122.

— 1952: Die Wirkung von Adenosintriphosphat, Muskeladenylsäure, Hefeadenylsäure und Adenosin auf die Teilung des Seeigeleis bei Anwendung von Hemmungsfaktoren. Exptl. Cell Research **3**, 584—596.

— and J. Runnström, 1951: Some Effects of Adenosine Triphosphate on the Cytoplasmic State, Division, and Development of the Sea Urchin Egg. Trans. N. Y. Acad. Sci. (Ser. II) **13**, 162—164.

— — 1952: Some Observations on the Effect of Colchicine and Adenosinetriphosphate on the Early Development of the Sea Urchin Egg. Arkiv Zool. (Ser. 2) **4**, 143—152.

Laird, A. K., O. Nygaard, H. Ris, and A. D. Barton, 1953: Separation of Mitochondria into two Morphological and Biological Different Types. Exptl. Cell Research **5**, 147—160.

de Lamirande, G., C. Allard, and A. Cantero, 1953: Electrophoretic Analysis of the Soluble Proteins of Cell Fractions Isolated from Regenerating Rat Liver, Liver Tumor, and from Liver of Rats Fed *p*-Dimethylaminoazobenzene. Cancer **6**, 179—183.

Lardy, H. A., 1952: Concerning the Mode of Action of Hormones. Résumés des Communications. IIe Congrès International de Biochimie. Paris. 56.

— J. H. Copenhaver, and H. Wellman, 1951: Discussion. Phosphorus Metabolism **1**, 587.

— and H. Wellman, 1952: Oxidative Phosphorylations: Rôle of Inorganic Phosphate and Aceptor Systems in Control of Metabolic Rates. J. biol. Chem. (Am.) **195**, 215—224.

Lavin, G. I., and A. W. Pollister, 1942: Some Observations with a Simplified Quartz Microscope. Biol. Bull. **83**, 299.

Lazarow, A., 1943: The Chemical Structure of Cytoplasm as Investigated in Professor Bensley's Laboratory During the Past 10 Years. Biol. Symposia **10**, 9—26.

— and S. J. Cooperstein, 1953: Studies on the Mechanism of Janus Green B Staining of Mitochondria. I. Review of the Literature. Exptl. Cell Research **5**, 56—69.

Lehninger. A. L., 1949: Esterification of Inorganic Phosphate Coupled to Electron Transport between Dihydrodiphosphopyridine Nucleotide and Oxygen. II. J. biol. Chem. (Am.) **178**, 625—644.
— 1951: Oxidative Phosphorylation in Diphosphopyridine Nucleotide-Linked Systems. Phosphorus Metabolism 1. 344—366.
— and S. W. Smith. 1949: Efficiency of Phosphorylation Coupled to Electron Transport between Dihydrodiphosphopyridine Nucleotide and Oxygen. J. biol. Chem. (Am.) **181**, 415—429.
Lenicque. P.. 1953: Étude sur l'évolution du chondriome au cours de la genèse du tissu musculaire et de sa dégénérescence provoquée par la colchicine. Arkiv Zool. **5**, 289—296.
Leuthardt. F.. 1952: Uréogénèse. cycle tricarboxylique et mitochondries. Symposium sur le Cycle Tricarboxylique. IIe Congrès International de Biochimie. Paris. 89—93.
— und E. Bujard. 1947: Über Glutaminbildung im Leber-Homogenat. Helvet. med. Acta **14**. 274—278.
— et J. Mauron. 1950: L'oxydation du puruvate dans des suspensions de mitochondries hépatiques. Helvet. physiol. et pharmacol. Acta **8**, 386—387.
— A. F. Müller. und H. Nielsen. 1949: Biologische Citrullinsynthese. Glutamin und α-Ureidoglutarsäure. Helvet. chim. Acta **32**. 744—756.
— et H. Nielsen. 1951: Recherches sur la synthèse biologique de l'acide hippurique. Helvet. chim. Acta **34**, 1618—1631.
— — 1952: Phosphorylation biologique de la thiamine. Helvet. chim. Acta **35**. 1196—1209.
— und E. Testa. 1951: Die Phosphorylierung der Fructose in der Leber. Helvet. chim. Acta **34**. 931—938.
Levine. C.. and E. Chargaff. 1952: Phosphatide Composition in Different Liver Cell Fractions. Exptl. Cell Research **3**. 154—162.
Lillie, R. D.. 1948: Histopathologic Technic. Philadelphia.
Lindahl. P. E.. 1936: Zur Kenntnis der physiologischen Grundlagen der Determination im Seeigelkeim. Stockholm.
— 1939: Über die biologische Sauerstoffaktivierung nach Versuchen mit Kohlenmonoxyd an Seeigeleiern und Keimen. Z.. vergl. Physiol. **27**. 136—168.
— and K. H. Kiessling. 1952: On Accumulation of Inorganic Pyrophosphate in the Cleaving Sea Urchin Egg Caused by Lithium Ions. Arkiv Kemi **3**. 97—105.
Lindberg. O.. 1951: Phosphorylation of Glyceraldehyde. Glyceric Acid and Dihydroxyacetone by Kidney Extracts. Biochim. et Biophys. Acta **7**, 349—353.
— and L. Ernster. 1952 a: On the Mechanism of Phosphorylative Energy Transfer in Mitochondria. Exptl. Cell Research **3**. 209—239.
— — 1951 b: On the Rôle of Adenine Nucleotides in Phosphorylative Energy Transfer in the Cyclophorase System. Résumés des Communications. IIe Congrès International de Biochimie. Paris. 36.
— M. Ljunggren. L. Ernster, and L. Révész. 1953: Isolation and some Enzymic Properties of Ehrlich Ascites Tumor Mitochondria. Exptl. Cell Research **4**. 243—245.
Lipmann, F.. 1939: An Analysis of the Pyruvic Acid Oxidation System. Cold Spring Harbor Symposia Quant. Biol. **7**. 248—259.
— 1940: A Phosphorylated Oxidation Product of Pyruvic Acid. J. biol. Chem. (Am.) **134**. 463—464.
— 1941: Metabolic Generation and Utilization of Phosphate Bond Energy. Advances in Enzymol. **1**. 99—162.
— 1945: Acetylation of Sulfanilamide by Liver Homogenates and Extracts. J. biol. Chem. (Am.) **160**. 173—190.
— 1946 a: Acetyl Phosphate. Advances in Enzymol. **6**. 231—267.
— in D. E. Green, 1946 b: Metabolic Process Patterns. Currents in Biochemical Research. New York. 137—148.
— M. E. Jones. and S. Black, 1952: Studies on the Mechanism of the ATP-CoA-Acetate Reaction Including a Survey of the Functions of CoA. Symposium sur le Cycle Tricarboxylique. IIe Congrès International de Biochimie. Paris. 55—63.
— — and R. M. Flynn. 1952: Enzymatic Phosphorylation of Coenzyme A by Adenosinetriphosphate. J. amer. chem. Soc. **74**. 2384—2385.
— N. O. Kaplan. G. D. Novelli, L. C. Tuttle. and B. M. Guirard. 1950: Isolation of Coenzyme A. J. biol. Chem. (Am.) **186**, 235—245.

Loomis, W. F., and F. Lipmann. 1948: Reversible Inhibition of the Coupling between Phosphorylation and Oxidation. J. biol. Chem. (Am.) **173**, 807—808.

Ludford, R. J., 1948: The Study of Living Malignant Cells by Phase Contrast and Ultra-Violet Microscopy. J. Roy. Microscop. Soc. **68**, 1—9.

Lwoff, A., 1950: Problems of Morphogenesis in Ciliates. New York.

Lynen, F., und R. Koenigsberger, 1951: Zum Mechanismus der Pasteurschen Reaktion: Der Phosphat-Kreislauf in der Hefe und seine Beeinflussung durch 2,4-Dinitrophenol. Liebigs Ann. Chem. **573**, 60—84.

— und E. Reichert, 1951: Zur chemischen Struktur der „aktivierten Essigsäure". Angew. Chem. **63**, 47—48.

— — und L. Rueff, 1951: Zum biologischen Abbau der Essigsäure. VI. „Aktivierte Essigsäure", ihre Isolierung aus Hefe und ihre chemische Natur. Ann. Chem. **574**, 1—32.

Mahler, H. R., A. Tomisek, and F. M. Huennekens, 1953: Studies on the Cyclophorase System. XXVI. The Lactic Oxidase. Exptl. Cell. Research **4**, 208—221.

Marshak, A., 1948: Evidence for a Nuclear Precursor of Ribo- and Desoxyribonucleic Acid. J. cellul. a. comp. Physiol. (Am.) **32**, 381—406.

— 1951: Purine and Pyrimidine Content of the Nucleic Acids of Nuclei and Cytoplasm. J. biol. Chem. (Am.) **189**, 607—613.

Marshall, A. J., 1952: The Structure of the So-called Golgi Body. Science Progr. **40**, 71—77.

Martius, C., 1952: Die ersten Reaktionen des Citronsäurezyklus. Symposium sur le Cycle Tricarboxylique. IIe Congrès International de Biochimie. Paris. 28—34.

— und F. Lynen, 1950: Probleme des Citronsäurezyklus. Advances in Enzymol. **10**, 167—222.

McEwen, W. K., and J. J. Eiler, 1950: Enzymatic Anodic Phosphorylation in Rat Brain Homogenates. Federation Proc. **9**, 200—201.

McGilvery, R. W., and Ph. P. Cohen, 1950: Enzymatic Synthesis of Ornithuric Acids. J. biol. Chem. (Am.) **183**, 179—189.

de Meio, R. H., and L. Tkacz, 1952: Conjugation of Phenol by Rat Liver Slices and Homogenates. J. biol. Chem. (Am.) **195**, 175—184.

Meves, F., 1908: Die Chondriosomen als Träger erblicher Anlagen. Cytologische Studien am Hühnerembryo. Arch. mikrosk. Anat. u. Entw.gesch. **72**, 816—867.

Meyer, A., 1920: Morphologisch-physiologische Analyse der Zelle. Jena.

Michaelis, L., 1900: Die vitale Färbung, eine Darstellungsmethode der Zellgranula. Arch. mikrosk. Anat. u. Entw.gesch. **55**, 558—575.

Mider, G. B., 1952: Some Aspects of Nitrogen and Energy Metabolism in Cancerous Subjects: A Review. Cancer Research **11**, 821—829.

Millican, R. C., S. M. Rosenthal, and H. Tabor, 1949: On the Metabolism of Histamine. I. Urinary Excretion Following Oral Administration. II. Conjugation in vitro. J. Pharmacol. (Am.) **97**, 4—13.

Monné, L., 1948: Functioning of the Cytoplasm. Advances in Enzymol. **8**, 1—69.

Monod, J., 1948: Facteurs génétiques et facteurs chimiques spécifiques dans la synthèse des enzymes bactériens. VIII. Unités Biologiques Douées de Continuité Génétique. Colloq. intern. centre natl. recherche sci. Paris. 181—199.

Moog, F., 1946: The Physiological Significance of the Phosphomonoesterases. Biol. Rev. Cambridge philos. Soc. **21**, 41—59.

Mudge, G. H., and J. V. Taggart, 1950 a: Effect of 2,4-Dinitrophenol on Renal Transport Mechanisms in Dog. Amer. J. Physiol. **161**, 173—180.

— — 1950 b: Effect of Acetate on Renal Excretion of p-Aminohippurate in Dog. Amer. J. Physiol. **161**, 191—197.

Muntwyler, E., S. Seifter, and D. M. Harkness, 1950: Some Effects of Restrictions of Dietary Protein on the Intracellular Components of Liver. J. biol. Chem. (Am.) **184**, 181—190.

Mühlethaler, K., A. F. Müller, und H. U. Zollinger, 1950: Zur Morphologie der Mitochondrien. Experientia **6**, 16—17.

Müller, A. F., und F. Leuthardt, 1949: Oxydative Phosphorylierung und Citrullinsynthese in den Lebermitochondrien. Helvet. chim. Acta **32**, 2349—2356.

Nachmansohn, D., 1951: Energy Sources of Bioelectricity. Phosphorus Metabolism **1**, 568—585.

— and A. L. Machado, 1943: The Formation of Acetyl Choline. A New Enzyme: "Choline Acetylase." J. Neurophysiol. **6**, 397—403.

— and I. B. Wilson, 1951: The Enzymic Hydrolysis and Synthesis of Acetylcholine. Advances in Enzymol. **12**, 259—339.

Nakada, H. I., and S. Weinhouse, 1950: Oxidation of Aspartic Acid by Rat Liver. J. biol. Chem. (Am.) **187**, 663—672.

Negelein, E., und H. Brömel, 1939: Isolierung eines reversibeln Zwischenprodukts der Gärung. Biochem. Z. **301**, 135—136.

Novelli, G. D., 1951: The Structure of Coenzyme A. Phosphorus Metabolism 1, 414—417.

— J. D. Gregory, R. M. Flynn, and F. J. Schmetz, 1951: Structure of Coenzyme A. Federation Proc. **10**, 229—230.

— N. O. Kaplan, and F. Lipmann, 1949: The Liberation of Pantothenic Acid from Coenzyme A. J. biol. Chem. (Am.) **177**, 97—107.

— and F. Lipmann, 1950: The Catalytic Function of Coenzyme A in Citric Acid Synthesis. J. biol. Chem. (Am.) **182**, 213—228.

Novikoff, A. B., E. Podber, and J. Ryan, 1950: Intracellular Distribution of Phosphatase Activity in Rat Liver. Federation Proc. **9**, 210.

Ochoa, S., 1943: Efficiency of Aerobic Phosphorylation in Cell-free Heart Extracts. J. biol. Chem. (Am.) **151**, 493—505.

— 1947: Chemical Processes of Oxidative Recovery. Ann. N. Y. Acad. Sci. **47**, 835—845.

— 1951: Biological Mechanisms of Carboxylation and Decarboxylation. Physiol. Rev. (Am.) **31**, 56—106.

— 1952: Enzymatic Synthesis of Citric Acid and other Reactions of the Tricarboxylic Acid Cycle. Symposium sur le Cycle Tricarboxylique. IIe Congrès International de Biochimie. Paris. 73—88.

— J. R. Stern, and W. C. Schneider, 1951: Enzymatic Synthesis of Citric Acid. II. Crystalline Condensing Enzyme. J. biol. Chem. (Am.) **193**, 691—702.

O'Kane, D. J., and I. C. Gunsalus, 1947: Accessory Factor Requirement for Pyruvate Oxidation. J. Bacter. (Am.) **54**, 20—21.

— — 1948: Pyruvic Acid Metabolism. A Factor Required for Oxidation by *streptococcus faecalis*. J. Bacter. (Am.) **56**, 499—506.

Opie, E. L., and G. I. Lavin, 1946: Localization of Ribonucleic Acid in the Cytoplasm of Liver Cells. J. exper. med. **84**, 107—112.

Palade, G. E., 1952: The Fine Structure of Mitochondria. Anat. Record **114**, 427—451.

Paul, M. H., M. Fuld, and E. Sperling, 1952: Cyclophorase System. XXII. Cyclophorase System of Rabbit Heart Muscle. Proc. Soc. exper. Biol. a. Med. (Am.) **79**, 349—352.

— and E. Sperling, 1952: Cyclophorase System. XXIII. Correlation of Cyclophorase Activity and Mitochondrial Density in Striated Muscle. Proc. Soc. exper. Biol. a. Med. (Am.) **79**, 352—354.

Peiss, C. N., and J. Field, 1948: A Comparison of the Influence of 2,4-Dinitrophenol on the Oxygen Consumption of Rat Brain Slices and Homogenates. J. biol. Chem. (Am.) **175**, 49—56.

Peters, R. A., 1952 a: Inhibitors of the Tricarboxylic Acid Cycle. Symposia sur le Cycle Tricarboxylique. IIe Congrès International de Biochimie. Paris. 64—72.

— 1952 b: Lethal Synthesis. Proc. Roy. Soc. London. B. **139**, 143—170.

Plaut, G. W. E., and K. A. Plaut, 1952: Oxidative Metabolism of Heart Mitochondria. J. biol. Chem. (Am.) **199**, 141—151.

Porter, K. R., A. Claude, and E. F. Fullam, 1945: A Study of Tissue Culture Cells by Electron Microscopy. J. exper. Med. (Am.) **81**, 233—246.

Potter, V. R., 1950: Biological Oxidations. Ann. Rev. Biochem. **19**, 1—20.

— and H. Busch, 1950: Citric Acid Content of Normal and Tumor Tissues in vivo Following Injection of Fluoroacetate. Cancer Research **10**, 353—356.

— and C. A. Elvehjem, 1936: A Modified Method for the Study of Tissue Oxidation. J. biol. Chem. (Am.) **114**, 495—504.

— and C. Heidelberger, 1950: Alternative Metabolic Pathways. Physiol. Rev. (Am.) **30**, 487—512.

— and G. G. Lyle, 1951: Oxidative Phosphorylation in Homogenates of Normal and Tumor Tissues. Cancer Research **11**, 355—360.

— — and W. C. Schneider, 1951: Oxidative Phosphorylation in Whole Homogenates and in Cell Particles. J. biol. Chem. (Am.) **190**, 293—301.

— and R. O. Recknagel, 1951: The Regulation of the Rate of Oxidation in Rat Liver Mitochondria. Phosphorus Metabolism 1, 377—391.

POTTER, V. R., R. O. RECKNAGEL, and R. B. HURLBERT, 1951: Intracellular Enzyme Distribution, Interpretations and Significance. Federation Proc. 10, 646—653.

— and A. E. REIF, 1951: Inhibition of an Electron Transport Component by Antimycin A. J. biol. Chem. (Am.) 194, 287—297.

PRESSMAN, B. C., and H. A. LARDY, 1952: Influence of Potassium and other Alkali Cations on Respiration of Mitochondria. J. biol. Chem. (Am.) 197, 547—556.

PRICE, J. M., E. C. MILLER, J. A. MILLER, and G. M. WEBER, 1949 a: Studies on the Intracellular Composition of Livers from Rats Fed Various Amoniazo Dyes. Cancer Research 9, 398—402.

— — — — 1949 b: Studies on the Intracellular Composition of Liver and Liver Tumor from Rats Fed 4-Dimethylaminoazobenzene. Cancer Research 9, 96—102.

— — — — 1950: Studies on the Intracellular Composition of Liver from Rats Fed Various Aminoazo Dyes. II. Cancer Research 10, 18—27.

RAAFLAUB, J., 1952: Die Korrelation zwischen Struktur und Aktivität von isolierten Leberzellmitochondrien. Helvet. physiol. Acta 10, 22—24.

RACKER, E., 1947: Spectrophotometric Measurement of Hexokinase and Phosphohexokinase Activity. J. biol. Chem. (Am.) 167, 843—854.

— 1951: The Mechanism of Action of Glyoxalase. J. biol. Chem. (Am.) 190, 685—696.

— and J. KRIMSKY, 1952: The Mechanism of Oxidation of Aldehydes by Glyceraldehyde-3-Phosphate Dehydrogenase. J. biol. Chem. (Am.) 198, 731—743.

RADIN, N. S., D. RITTENBERG, and D. SHEMIN, 1950 a: The Rôle of Glycine in the Biosynthesis of Heme. J. biol. Chem. (Am.) 184, 745—753.

— — — 1950 b: The Rôle of Acetic Acid in the Biosynthesis of Heme. J. biol. Chem. (Am.) 184, 755—767.

RATNER, S., 1951: Urea Formation. Phosphorus Metabolism 1, 601—619.

RECKNAGEL, R. O., and V. R. POTTER, 1951: Mechanism of the Ketogenic Effect of Ammonium Chloride. J. biol. Chem. (Am.) 191, 263—275.

REED, L. J., and B. G. DE BUSK, 1952 a: Lipothiamide and its Relation to a Thiamin Coenzyme Required for Oxidative Decarboxylation of α-Ketoacids. J. amer. chem. Soc. 74, 3457.

— — 1952 b: Lipothiamide Pyrophosphate: Coenzyme for Oxidative Decarboxylation of α-Ketoacids. J. amer. chem. Soc. 74, 3964—3965.

— — I. C. GUNSALUS, and C. S. HORNBERGER, 1951: Crystalline α-Lipoic Acid: A Catalytic Agent Associated with Pyruvate Dehydrogenase. Science 114, 93—94.

REGAUD, C., 1909: Participation du chondriome à la formation des graines de ségrégation dans les cellules des tubes contournes des rein (chez les ophidiens et les amphibiens). C. r. Soc. Biol. 66, 1034—1036.

RITCHIE, D., and P. HAZELTINE, 1953: Mitochondria in Allomyses under Experimental Conditions. Expetl. Cell Research 5, 261—274.

DE ROBERTIS, E. D. P., W. W. NOWINSKI, and F. A. SAEZ, 1948: General Cytology. Philadelphia and London.

ROBINSON, H. W., and C. G. HOGDEN, 1940: The Biuret Reaction in the Determination of Serum Proteins. I. A Study of the Conditions Necessary for the Production of a Stable Color which Bears a Quantitative Relationship to the Protein Concentration. J. biol. Chem. (Am.) 135, 707—725.

ROJAS, P., and DE ROBERTIS, 1936: Contribución al estudio de la citología del eritocito del Bufo arenarum Hensel. Rev. Soc. argent. Biol. 12, 325.

RONZONI, E., and E. EHRENFAST, 1936: The Effect of Dinitrophenol on the Metabolism of Frog Muscle. J. biol. Chem. (Am.) 115, 749—768.

ROSENBERG, TH., and W. WILBRANDT, in G. H. BOURNE, and J. F. DANIELLI, 1952: Enzymatic Processes in Cell Membrane Penetration. Cytology 1, 65—92.

RUNNSTRÖM, J., 1952: The Cytoplasm. Its Structure and Rôle in Metabolism. Growth and Differentiation. Modern Trends in Physiol. and Biochem. New York. 47—76.

— and G. KRISZAT, 1950 a: On the Effect of Adenosine Triphosphoric Acid and of Ca on the Cytoplasm of the Egg of the Sea Urchin, Psammechinus miliaris. Exptl. Cell Research 1, 284—303.

— — 1950 b: On the Influence of ATP of the Fertilization and Segmentation of the Sea Urchin Egg, Strongylocentrotus lividus. Exptl. Cell Research 1, 497—499.

Sacktor. B., 1953: Investigations on the Mitochondria of the House Fly, *Musca domestica* L. J. gen. Physiol. (Am.) **36**, 371—387.

Sakami. W., 1948: The Conversion of Formate and Glycine to Serine and Glycogen in the Intact Rat. J. biol. Chem. (Am.) **176**, 995—996.

Sanadi, D. R., and J. W. Littlefield. 1951: Studies on α-Ketoglutaric Oxidase. I. Formation of "Active" Succinate. J. biol. Chem. (Am.) **193**, 683—689.

— — 1952: Rôle of Coenzyme A and DPN in the Oxidation of α-Ketoglutaric Acid. Science **116**, 327—328.

— — and R. M. Bock. 1952: Studies on α-Ketoglutaric Oxidase. II. Purification and Properties. J. biol. Chem. (Am.) **197**. 851—862.

Schein. A. H., and E. Young, 1952: Intracellular Localization of Arginase in Homogenates of Rat Liver Suspended in Distilled Water. Exptl. Cell Research **3**, 383—387.

Schneider. W. C., 1946 a: Intracellular Distribution of Enzymes. I. The Distribution of Succinic Dehydrogenase, Cytochrome Oxidase, Adenosinetriphosphatase. and Phosphorus Compounds in Normal Rat Tissues. J. biol. Chem. (Am.) **165**, 585—593.

— 1946 b: Intracellular Distribution of Enzymes. II. The Distribution of Succinic Dehydrogenase. Cytochrome Oxidase, Adenosinetriphosphatase, and Phosphorus Compounds in Rat Liver and in Rat Hepatomas. Cancer Research **6**. 685—690.

— 1947: Nucleic Acids in Normal and Neoplastic Tissues. Cold Spring Harbor Symposia Quant. Biol. **12**, 169—178.

— 1948: Intracellular Distribution of Enzymes. III. The Oxidation of Octanoic Acid by Rat Liver Fractions. J. biol. Chem. (Am.) **176**, 259—266.

— and G. H. Hogeboom, 1950: Intracellular Distribution of Enzymes. V. Further Studies on the Distribution of Cytochrome *c* in Rat Liver Homogenates. J. biol. Chem. (Am.) **183**, 123—128.

— — 1951: Cytochemical Studies of Mammalian Tissues: the Isolation of Cell Components by Differential Centrifugation: A Review. Cancer Research **11**, 1—22.

— — and V. R. Potter, 1949: Intracellular Distribution of Enzymes. IV. The Distribution of Oxalacetic Oxidase Activity in Rat Liver and Rat Kidney Fractions. J. biol. Chem. (Am.) **177**, 893—903.

Schweet. R. S., and K. Cheslock. 1952: Pyruvic Oxidase of Pigeon Breast Muscle. III. Factors Influencing Enzymatic Activity. J. biol. Chem. (Am.) **199**, 749—756.

— M. Fuld. K. Cheslock, and M. H. Paul, 1951: Initial Stages of Pyruvate Oxidation. Phosphorus Metabolism **1**, 246—259.

— B. Katchman, R. M. Bock, and V. Jagannathan, 1952: Pyruvic Oxidase of Pigeon Breast Muscle. II. Physicochemical Studies. J. biol. Chem. (Am.) **196**, 563—567.

Schweigert, B. S., B. T. Guthneck, J. M. Price, J. A. Miller, and E. C. Miller, 1949: Amino Acid Composition of Morphological Fractions of Rat Livers and Induced Liver Tumors. Proc. Soc. exper. Biol. a. Med. (Am.) **72**, 495—501.

Shannon, J. A., and S. Fisher, 1938: The Renal Tubular Reabsorption of Glucose in the Normal Dog. Amer. J. Physiol. **122**, 765—774.

Shaver, J. R.. and J. Brachet, 1949: The Exposition of Chorioallantoic Membranes of the Chick Embryo to Granules from Embryonic Tissue. Experientia **5**, 235.

Shemin, D., and S. Kumin, 1952: The Mechanism of Porphyrin Formation. The Formation of a Succinyl Intermediate from Succinate. J. biol. Chem. (Am.) **198**, 827—837.

— and J. Wittenberg, 1951: The Mechanism of Porphyrin Formation. The Rôle of the Tricarboxylic Acid Cycle. J. biol. Chem. (Am.) **192**, 315—334.

Siekevitz, Ph., 1952: Uptake of Radioactive Alanine in vitro into the Proteins of Rat Liver Fractions. J. biol. Chem. (Am.) **195**, 549—565.

— and D. M. Greenberg, 1949: The Biological Formation of Serine from Glycine. J. biol. Chem. (Am.) **180**. 845—856.

— and Potter, V. R., 1953: The adenylate kinase of rat liver mitochondria. J. biol. Chem. (Am.) **200**, 187—196.

Sjöstrand, F. S., 1951: A Method for Making Ultra-Thin Tissue Sections for Electron Microscopy at High Resolution. Nature **168**, 646—647.

— 1953 a: Electron Microscopy of Mitochondria and Cytoplasmic Double Membranes. Nature **171**. 30—32.

SJÖSTRAND, F. S., 1953 b: The Ultrastructure of the Outer Segments of Rods and Cones of the Eye. As Revealed by Electron Microscope. J. cellul. a. comp. Physiol. (Am.) 42, 15—44.

— and J. RHODIN, 1953: The Ultrastructure of the Proxymal Convoluted Tubules of the Mouse Kidney as Revealed by High Resolution Electron Microscopy. Exptl. Cell Research 4, 426—456.

SLATER, E. C., 1948: The Measurement of the Cytochrome Oxidase Activity of Enzyme Preparations. Biochem. J. 44, 305—318.

— 1949 a: A · Comparative Study of the Succinic Dehydrogenase-Cytochrome System in Heart Muscle and in Kidney. Biochem. J. 45, 1—8.

— 1949 b: The Action of Inhibitors on the System of Enzymes which Catalyse the Aerobic Oxidation of Succinate. Biochem. J. 45, 8—13.

— 1949 c: A Respiratory Catalyst Required for the Reduction of Cytochrome c by Cytochrome b. Biochem. J. 45, 14—30.

— 1950 a: The Components of the Dihydrocozymase Oxidase System. Biochem. J. 46, 484—499.

— 1950 b: The Dihydrocozymase-Cytochrome c Reductase Activity of Heart Muscle Preparation. Biochem. J. 46, 499—503.

— 1950 c: Phosphorylation Coupled with the Reduction of Cytochrome c by α-Ketoglutarate in Heart Muscle Granules. Nature 166, 982—984.

— and K. W. CLELAND, 1952: Stabilization of Oxidative Phosphorylation in Heart-Muscle Sarcosomes. Nature 170, 118—119.

— and F. A. HOLTON, 1953: The Adeninenucleotide Specificity of Oxidative Phosphorylation. Biochem. J. 53, II.

SNELL, E. E., G. M. BROWN, V. J. PETERS, J. A. CRAIG, E. L. WITTLE, J. A. MOORE, V. M. McGLOHON, and O. D. BIRD, 1950: Chemical Nature and Synthesis of the Lactobacillus Butgariens Factor. J. amer. chem. Soc. 72, 3349—3350.

SNELLMAN, O., and C. E. DANIELSON: The Experimental Study of the Biosynthesis of the Reserve Globulin in Pea Seeds. Exptl. Cell Research (in press).

— and B. GELOTTE, 1951: The Prothetic Group of Actin. Nature 168. 462.

SNOKE, J. E., and F. ROTHMAN, 1951: Glutathione Synthesis from Glutamyl Cysteine and Glycine. Federation Proc. 10, 249.

SONNEBORN, T. M., 1951: Some Current Problems on Genetics in the Light of Investigations on *Chlamydomonas* and *Paramecium*. Cold Spring Harbor Symposia Quant. Biol. 16, 483—503.

SOODAK, M., and F. LIPMANN, 1948: Enzymatic Condensation of Acetate to Acetoacetate on Liver Extracts. J. biol. Chem. (Am.) 175, 999—1000.

SPECK, J. F., 1949: The Enzymatic Synthesis of Glutamine, a Reaction Utilizing Adenosine Triphosphate. J. biol. Chem. (Am.) 179, 1405—1426.

STADTMAN, E. R., G. D. NOVELLI, and F. LIPMANN, 1951: Coenzyme A Function in and Acetyl Transfer by the Phosphotransacetylase System. J. biol. Chem. (Am.) 191, 365—376.

STERN, H., V. ALLFREY, A. E. MIRSKY, and H. SAETREN. 1952: Some Enzymes of Isolated Nuclei. J. gen. Physiol. (Am.) 35. 559—578.

STERN, J. R., and S. OCHOA, 1951: Enzymatic Synthesis of Citric Acid. I. Synthesis with Soluble Enzymes. J. biol. Chem. (Am.) 191. 161—172.

— — and F. LYNEN, 1952: Enzymatic Synthesis of Citric Acid. V. Reaction of Acetyl Coenzyme A. J. biol. Chem. (Am.) 198. 313—321.

— B. SHAPIRO, E. R. STADTMAN, and S. OCHOA, 1951: Enzymatic Synthesis of Citric Acid. III. Reversibility and Mechanism. J. biol. Chem. (Am.) 193. 703—720.

STILL, J. L., M. V. BUELL, and D. E. GREEN, 1950 a: Studies on the Cyclophorase System. VIII. Oxidation of l-Glutamate. Arch. Biochem. 26. 406—412.

— — — 1950 b: Studies on the Cyclophorase System. IX. Oxidation of l-Alanine. Arch. Biochem. 26, 413—419.

— — W. E. KNOX, and D. E. GREEN, 1949: Studies on the Cyclophorase System. VII. d-Aspartic Oxidase. J. biol. Chem. (Am.) 179. 831—837.

SWENDSEID, E., F. H. BETHELL, W. W. ACKERMANN. 1951: The Intracellular Distribution of Vitamin B_{12} and Folinic Acid in Mouse Liver. J. biol. Chem. (Am.) 190. 791—798.

SZORENYI, E. T., and O. P. CHEPINOGA, 1946: Ukrain. Biokhim. Zhur. 18. 137.

Taggart, J. V., and R. P. Forster, 1950: Renal Tubular Transport Effect of 2, 4-Dinitrophenol and Related Compounds on Phenol Red Transport in the Isolated Tubules of the Flounder. Amer. J. Physiol. **161**, 167—172.

— and R. B. Krakaur, 1949: Studies on the Cyclophorase System. V. The Oxidation of Proline and Hydroxyproline. J. biol. Chem. (Am.) **177**, 641—653.

Teply, L. J., 1949: Studies on the Cyclophorase System. XIV. Mechanism of Action of 2, 4-Dinitrophenol. Arch. Biochem. **24**, 383—388.

Umbreit, W. W., R. H. Burris, and J. F. Stauffer, 1947: Manometric Techniques and Related Methods for the Study of Tissue Metabolism. Minneapolis.

Utter, M. F., 1951: Adenosine Triphosphate and Carbon Dioxide Fixation. Phosphorus Metabolism 1, 646—657.

— and H. G. Wood, 1946: The Fixation of Carbon Dioxide in Oxalacetate by Pigeon Liver. J. biol. Chem. (Am.) **164**, 455—476.

— — 1951: Mechanisms of Fixation of Carbon-Dioxide by Heterotrophs and Autotrophs. Advances in Enzymol. **12**, 41—151.

Vendrely, C., 1950: Contribution à l'étude cytochimique des acides nucléiques de quelques organites cellulaires. Arch. Anat. etc. (Fr.) **33**, 81—112.

Vennesland, B., E. A. Evans, and K. I. Altman, 1947: The Effects of Triphosphopyridine Nucleotide and of Adenosine Triphosphate on Pigeon Liver Oxalacetic Carboxylase. J. biol. Chem. (Am.) **171**, 675—686.

Waelsch, H., 1952: Certain Aspects of Intermediary Metabolism of Glutamine, Asparagine, and Glutathione. Advances in Enzymol. **13**, 237—319.

Walaas, O., E. Wallaas, and F. Løken, 1952: Effect of Oestradiol Monobenzoate on the Metabolism of Rat Uterine Muscle. Acta Endocrinol. **10**, 201—211.

Warburg, O., 1908: Beobachtungen über die Oxydationsprozesse im Seeigelei. Hoppe-Seylers Z. **57**, 1—16.

— 1910: Über die Oxydationen in lebenden Zellen nach Versuchen an Seeigelei. Hoppe-Seylers Z. **66**, 305—340.

— 1913: Über sauerstoffatmende Körnchen aus Leberzellen und über Sauerstoffatmung in Berkefeldfiltraten wäßriger Leberextrakte. Pflügers Arch. **154**, 599—617.

— 1926: Über die Wirkung von Blausäureäthylester (Äthylcarbylamin) auf die Pasteursche Reaktion. Biochem. Z. **172**, 432—441.

— und W. Christian, 1939: Isolierung und Kristallisation des Proteins des oxydierenden Gärungsferments. Biochem. Z. **303**, 40—68.

Watanabe, M. I., and C. M. Williams, 1951: Mitochondria in the Flight Muscles of Insects. I. Chemical Composition and Enzymatic Content. J. gen. Physiol. (Am). **34**, 675—689.

Weinhouse, S., 1951: Studies on the Fate of Isotopically Labelled Metabolites in the Oxidative Metabolism of Tumors. Cancer Research 11, 585—591.

— R. H. Millington, and C. E. Wenner, 1951: Metabolism of Neoplastic Tissue. I. The Oxidation of Carbohydrate and Fatty Acids in Transplanted Tumors. Cancer Research 11, 845—850.

Wenner, C. E., M. A. Spirtes, and .S. Weinhouse, 1952: Metabolism of Neoplastic Tissue. II. A Survey of Enzymes of the Citric Acid Cycle in Transplanted Tumors. Cancer Research 12, 44—49.

— and S. Weinhouse, 1953: Metabolism of Neoplastic Tissue. III. Diphosphopyridine Nucleotide Requirements for Oxidations by Mitochondria of Neoplastic and Non-neoplastic Tissues. Cancer Research 13, 21—26.

Williams, J. N., 1951: Intracellular Distribution of Choline Oxidase. J. biol. Chem. (Am.) **194**, 139—142.

— 1952 a: A Study of the Rôle of High Energy Phosphate in Mitochondrial Choline Oxidase. J. biol. Chem. (Am.) **198**, 579—585.

— 1952 b: Studies of the Stability of Mitochondrial Choline Oxidase. J. biol. Chem. (Am.) **197**, 709—715.

Zamecnik, P. C., 1952: The Biochemistry of Neoplastic Tissue. Ann. Rev. Biochem. **21**, 411—430.

Zollinger, H. U., 1948: Cytologic Studies with the Phase Microscope. Amer. J. Path. **24**, 545—588.

— 1950: Les Mitochondries (Leur étude à l'aide du microscope à contraste de phases). Revue d'Hématologie 5, 696—745.

— 1951: Beitrag zur Frage der Mitochondrienregeneration. Verhandlungen der Deutschen Gesellschaft für Pathologie. Hannover. 125—128.